View from the Cockpit

Cover photo by T. L. Morgan

View from the Cockpit

by Len Morgan

For John Guber whose
enthusiasm helps keep history alive —
with warm regards —
Len Morgan

Sunflower University Press®
1531 YUMA • MANHATTAN, KANSAS 66402-4228, USA

Most of these chapters have been published previously in *Air Facts, Air Line Pilot*, and *FLYING*.

Library of Congress Cataloging in Publication Data

Morgan, Len.
View from the cockpit.

1. Morgan, Len. 2. Air pilots — United States — Biography.
I. Title.
TL540.M66A38 1985 629.13'092'4 [B] 85-12720
ISBN 0-89745-072-8 (pbk.)

Books by Len Morgan

The P-51 Mustang
The P-47 Thunderbolt
The Planes the Aces Flew (with R. P. Shannon)
The Douglas DC-3
The AT-6 Harvard
Airliners of the World
Fifty Famous Tanks (with George Bradford)
Crack Up!
Aviation Hall of Fame
The Boeing 727 Scrapbook (with T. L. Morgan)

For
Cris,
Bo,
Mark,
Bri,
and
Mo

Contents

Introduction

Flying. Who can explain its fascination? Not I. At one time I could, but that was long before I got near an airplane. Back then, flying promised adventure, drama, an escape from the lackluster existence my elders seemed to endure, a chance to be part of something new, different, and spectacular. I was quite sure of all that.

I was not disappointed. All the promises were fulfilled. Yet adventure, drama, and escape were not peculiar to flying. They could be found in many fields. And, flying soon evolved into a far more disciplined and restrictive pursuit than any of the dreary ground vocations I dreaded. Yet it still lost none of its appeal.

The challenge of learning new aircraft and new ways to fly them is part of it. There's the rare satisfaction of safely completing a flight which would have been impossible a short while ago. This, for me, has been the real pleasure of watching the science develop and flying skills improve as they have. My youthful unbounded confidence in the future of aviation was vindicated; it's nice to be proved right.

I believe most flying is useful, that aviation serves worthwhile purposes, and that those of us involved can be proud of what we do. We accomplish things. But this does not fully explain the fascination of flying.

The inherent stability of the flying life has something to do with it. Much has been made of technical advance. A new design is obsolete before it leaves the drawing board and is soon replaced by another which is larger, faster, more efficient. The basics of flying never change, however, nor can they be changed. Today's latest marvel lifts and flies in the same way and for the same reasons the Wrights flew at Kitty Hawk in 1903. The good practices that kept pilots alive in the beginning preserve them now. When repeated, the mistakes of the dim past have identical consequence. The inflexible laws of flying are well known. A pilot can stake his life on them. In fact, he must.

Flying's rewards must be paid for. The price is often stiff. There is nothing adventuresome about study and hard work, yet successful completion of a tough course produces pride in achievement. You get what you pay for and there is an appeal in enjoying something for which you have dearly paid.

Flying plays on the emotions. It can be frustrating, fatiguing, terrifying — and most delightful. The pilot who has not been completely exasperated, dead on his feet, frightened witless, or enthralled beyond words has few hours in his log. His times will come. There is something to be said for work that leaves vivid recollections.

What a swirl of impressions are left with every airman, no matter what he flew or where he went! I set out to see the world, naively supposing that one could see it all in a lifetime. Thanks to flying, I saw some of it, enough to discard any notions I otherwise would have carried to the grave. From above, the vista is much the same wherever you go.

The patterns of land and water, of plains, mountains, deserts, lakes, and rivers repeat themselves. The boundary between Libya and Egypt is no more apparent than the one between us and Canada. A snapshot of the Rockies may later be mistaken for one made over Switzerland or Peru, not that it really matters. I have color slides of a memorable sunrise over the South China Sea which recall a hundred other dawns in a hundred other places. A shot of the Intertropical Convergence near the Equator could be passed off as a Kansas line squall.

Looking down on Flanders or Inchon or Manassas you have to wonder why this or that fragment of earth was once so bitterly contested. You encompass in a glance an area where 18,000 men exchanged their lives for five miles of dismal swamp during a single day. The forgotten shell holes survive.

The Pyramids of Giza, the Panama Canal, and Boulder Dam assume new meaning when seen from the air, as do the pioneers' wagon tracks still visible in our West. Like a large oil painting, earth is best viewed from a distance. From above, the coarse brush strokes of man's desecrations are reasonable and even seem to enhance the natural background. Manhattan's glistening towers now rise through the morning mists like a divinely inspired tribute. London's twisting streets flank the Thames as though laid out in the beginning.

Within flying minutes of every congested city is a lovely countryside, in comparison sparcely populated, a region of tranquility, a place with elbow room, yet, down on the highways, tiny cars race from one concrete jungle to the next, their riders seemingly unaware of their restful surroundings, or perhaps uncomfortable in them, anxious to rejoin the throng. Flying poses more questions than it answers.

Now retired, I truly miss the cockpit view, even though it most often meant seeing things and places I'd seen a thousand times before.

There was something about starting out once again along a well known route to a familiar destination that I will always miss. The chance to practice skills, to do something I thoroughly enjoyed, was only part of it. An explanation for the rest remains elusive.

Once I flew copilot for a veteran finishing up his last week of a 37-year airline career. He soloed in the early 1920s, dusted cotton, barnstormed, and flew right seat in Fords with another line before finding a home with us. He was a consummate pilot with 34,000 hours logged without scratching an airplane. He was also a great fellow to fly with.

As we took position on the runway and waited, looking down the 10,000 feet of concrete ahead, I wondered how many times before he had done it. What was he thinking now?

I wanted to ask, but didn't. Perhaps he read my thoughts.

"You know, he said, "this is just as much fun now as it ever was. Can you understand that?"

I could.

These following chapters, then, make an indirect attempt at an explanation — or perhaps a clarification — of my *view from the cockpit*.

Len Morgan

1.

Sea + Rail = Air

The first pilot I ever met was the one who taught me to fly. Why I was there in the first place I cannot explain to this day. Almost no one then was "air-minded," least of all my parents. But Dad and Mother were travel-minded. Displaced Britons, they had traveled widely by the measures of those times and gave their sons the itch to go places and see things. Despite his minuscule minister's salary, Dad somehow managed a family trip to the old country in 1927 and again in 1933, experiences still vivid in my recollections.

He was an avid rail and ship fan. The *Cheltenham Flyer, Flying Scotsman,* and *Cornish Riviera* were typical topics of dinner talk, these being crack British trains. Or it was the *Berengaria, Britannic,* or *Aquitania,* great British liners. He had ridden the trains and many of the ships.

We learned the basics early, that White Star Line vessels bore names ending in *ic* – *Baltic, Homeric, Celtic* and so on, while Cunard (pronounced *Cue* nard, not *Coo* nard, mind you) used *ia,* as in *Laconia, Carpathia,* and *Caronia.* A militant nationalist in maritime matters ("Look at the funnels on the *Normandie!* She's got no lines at all. Well, that's the French for you."), he conceded that Americans built splendid railroad locomotives.

Robert Louis Stevenson wrote, "I travel not to go anywhere, but to go. I travel for travel's sake. The great affair is to go." Dad would have put it, "I travel to watch the engine at work, the great ship under way." In daydreams I went them both one better: to sit in the cab, hand on throttle, or stand on the liner's lofty bridge surely would be greater thrills and the more I read about flying the more I wondered if being strapped into a cockpit might not be the ultimate adventure.

Dad imbued me with a fascination with railroading and sea travel I have never lost. Reading *Popular Aviation* and *Air Trails* seemed almost disloyal, yet flying had an appeal impossible to dismiss. There was something about it. I could not explain it then and I cannot now.

In 1929 we were in New York. In the main concourse of Pennsylvania Station was a glistening Ford trimotor propped up in flying position, an ad for the new transcontinental Air Transport. I looked in at the wicker seats, the cluttered cockpit, and ran my fingers along the cool corrugated skin, absolutely enthralled. Dad shook his head in wonder but said nothing.

We moved to Louisville, which boasted the state's largest airport, Bowman Field, a square mile of grass with two hangars. Half a dozen airliners came and went every day — Douglas DC-2s and, later, huge DC-3s. Dad and I often stood behind the fence on a summer evening, watching American Airlines' *Flagship Arizona*, with admiral's pennants aflutter above cockpit windows, roll to a stop beneath the floodlights, its big Wrights expiring with a death rattle of slowing gears. No true disciple of transportation could be unmoved. Dad was intrigued but unconvinced. "If it was a matter of life and death, I'd try it." He went that far.

While I haunted the airport, a younger brother set up camp beside the L&N Railroad tracks which ran a block from the manse. Somehow I raised four dollars and bought us a hop in a gull-wing Stinson *Reliant*. It was worth the risk of paternal wrath to make David see the light but, alas, he was already permanently unsettled by soft coal smoke and has for 40 years edited the leading railroad magazine. However, that ride may account for his hobby of collecting airlines, airliners, and airports. He's been ticketed in more than 50 types from Ford trimotor to *Concorde* on as many airlines to 175 places around the globe. Our kid brother went in another direction but has been a lifelong enthusiast of sorts. Mick would not cross town to watch the Thunderbirds, but would cross the state to watch a World War II P-40 fly, or just sit there with its V-12 Allison idling.

Mother and Dad eventually flew, first to England in a Pan Am *Constellation* with a fuel stop at Gander, later in an SAS DC-6. We took them to Chicago, Denver, Seattle, and other domestic points on our airline, and they flew to Texas to visit us many times. Mother, who had always regarded the destination as the sole justification of any trip (she could abide a Pullman but was a poor sailor prone to queasiness before the gangplank was away), loved air travel. The perfect passenger, she had implicit faith in crew and machine. She enjoyed flying for its own sake and was sorry to see it end. We bought an old four-seat Fairchild 24 and she begged a ride. She took the stick and kept us straight and level for several minutes, carefully listening to my explanation of instrument readings. "If I were younger, I could learn to do this right," she said. She would have been a good pilot.

Dad was not so sure. Aboard airline flights he watched everything, listened intently to all announcements, asked innumerable questions, and, as many British passengers do, recorded the details of each trip. "That's the same Delta DC-8 we rode over here last March," he'd say. But he never relaxed and enjoyed flying. Mother said he insisted on an aisle seat and never looked out during takeoff or landing. I think he prayed.

I fixed that. We took them on our nonstop 747 flight from Dallas to Honolulu, seeing to it he was installed in a window seat ahead of the wing. I promised to sit next to him and answer questions. Three hours into the trip we crossed San Francisco and I pointed out the bay and famous bridge. He gazed for a long while at the great wing and monstrous engines, then leaned back and said, "This is *wonderful!*" After the eight-hour run he said, "You know, I'd get right back on for the return trip." He saw the huge Boeing as a flying *Mauritania*. But he refused to ride in the Fairchild.

As passenger sea trade waned and the great ships he knew were retired and scrapped, he lost much of his interest. An outsider can understand his snort of derision at the *QE2*. He kept up with railroading here and in Britain, mostly reading about the age of steam and great trains of his youth. He considered the diesel a glorified streetcar.

One evening during his last year we sat among his books and framed pictures and models (which included one of our orange 747) and I listened again to a story from his boyhood. It happened in 1906 when he crossed with his father aboard the *Adriatic*. (Grandpa made 54 round trips by sea, 104 North Atlantic crossings, spending a year and a half as an ocean passenger.) He admired from a distance the ship's captain, a distinguished, bearded old salt in frock coat with gold stripes, and asked to meet him. It was arranged. The veteran sailor answered the boy's questions and signed his autograph book — "Edward J. Smith, Master, *R.M.S. Adriatic*." Captain John Smith was nearing the end of a long career. Six years later he reached the top of his profession and was given the highest honor the White Star Line could bestow, command of its newest jewel on her maiden voyage.

The *Titanic*.

2.

A Rare Breed

More than 40 pilots contributed to my professional education, teaching the basics and introducing new equipment. Without exception they were very good pilots, yet no more than 15 deserved the title, "instructor." The rest were check pilots, something else altogether. A check pilot tells you what you're doing wrong; an instructor tells you why, and how to do it right.

A mild-mannered Canadian of infinite patience got the first shot at me and may never have had less to work with. After a particularly exasperating hour of dual he said, "My grandmother could have done better," making me feel a foot tall. He worked me ragged, prodding, coaxing, kidding through the voice tube. Or saying nothing. Silence from the back seat spoke volumes. When eventually I heard, "That was much better," I was elated. He could find it if it was anywhere in you. He was an instructor.

My second mentor was also an instructor, though a completely different personality. He was also a brave young man to strap into the rear seat of a *Harvard* behind a kid with 60 hours total time. The *Harvard* could eat the lunch of a ham-fisted driver. He calmly restrained me from killing us both until I more or less got the hang of it. I owe those gentlemen. Had they been the verbal bullies who too often filled instructor seats in military trainers, I'd be thinking back now on 45 years of something not half the fun of flying.

The idea that flying skill is best imparted through a tirade of personal abuse goes way back. The terrible rage of impatient instructors is frequently mentioned in World War I accounts. Hollywood fostered this image during the 1930s with such epics as *West Point of the Air* (an overdone thing filmed at Randolph Field) with such success that my generation fully expected to be yelled at all through training. It was a real surprise that many of our mentors were warm, helpful, mature individuals.

There were plenty of the other sort, however, and there was no option but to try to learn from whomever you drew. Good or otherwise, an instructor held the power of life and death over his charge and could terminate a budding career virtually at his own whim. After graduation, an arrogant boor was an inexcusable obstacle to advancement. You learned to tune him out and teach yourself what had to be learned about a new machine which needlessly compounded the problem. Learning to fly a new plane is a devilishly difficult business without personality conflicts to worry about.

Immediately after the war, airline training schools included some instructors who should have remained out on the line. The worst was the fellow who, as an "essential war worker," had for whatever reasons sidestepped military service. His self-appointed mission was to whittle veterans down to size. "Now you're going to learn to do it the *airline* way" — that kind of nonsense, as though recklessness had been the military way. Again, you pretended to listen but went right ahead doing it the same way. The new hire with B-24 combat time in his log has little trouble copiloting a plodding DC-3 "the airline way."

Most airline know-how is acquired out on the line, not from books. Every captain is an instructor whether or not he cares to be, and many do not. The majority of our skippers taught by example, abetted with suggestion and comment, and from those grand fellows you learned the secrets of the trade. They commanded in the finest sense of the word. Others were narrow in their approach, insisting there was but one right way to do anything — their way, of course. A few regarded teaching as an obligation that went with captain's stripes. With them every hour was a training hour, a 10-year copilot receiving the same detailed critique as the flight engineer who upgraded last week. This minority labored under the delusion that all the brains sat in the left seat, and it was tiresome to fly with them.

From all I hear, the military is now considerably more selective in picking instructors. The airline picture certainly brightened over the years. Management finally came to see that the pompous tyrant who humiliated and frustrated his students was needlessly increasing training expense. Pilots themselves demanded a better shake. The working agreement of most carriers came to include a section on training. One proviso — that if a student did not hit it off with his instructor he could ask for another. And he could request extra airplane or flight simulator time for pre-checkride practice. The safety record proves the new approach works well.

Browbeating too often passed for instruction in the past. Little learning progress was made by needling an accomplished airman about his wandering speed and altitude. He could read the instruments and knew he was out of tolerance; what he wanted to know was why. A real instructor analyzed the problem and suggested a solution, remembering his own initial confustion and the corrections that worked for him. He was a blending of knowledge, patience, and firmness that quickly brought out the best in a student. He was a genuine teacher; few pilots could fill his shoes.

Out on the line where most of it was taught, the situation improved. The cockpit atmosphere became more cordial and cooperative than it had been 30 years earlier. Again, the steadily improving safety record reflects a change for the better. There are still left-seaters who emulate the old school. They were not favorites with copilots and engineers. A few are downright despised. Interestingly, their ranks are largely made up of individuals who, as copilots, protested the loudest when teamed with petty tyrants.

Which must prove something.

3.

Nerveless Men Only

"Anyone who has common sense and patience may learn to fly," wrote Francis Collins in 1917 in one of the few mentions of airman qualifications to be found in early aviation writing. While ballooning inspired hundreds of accounts and the first 15 years of airplanes (Kitty Hawk to the Armistice) produced an avalanche of books, little of that fascinating literature dwells on the aeronauts and pilots themselves. It concentrates on their exploits.

Then, as now, most writers were enthralled by the machine; few explored the mystique of piloting by taking the young gods apart to see what made them tick. Young readers must have yearned to learn what special attributes were demanded by the new art.

Well, it was a man's work—at least in the view of pilots. "Aviation is for grown men, alert, strong, sturdy and above all capable of endurance; such qualities are not often found in women," wrote French hero Andre Beaumont in 1912. "Flying is scarcely a suitable sport for women," agreed British airmen Gustav Hamel and Charles Turner in 1914, adding, "So far no woman has become a really good pilot." This was a bit strong, considering the achievements of certain ladies by that time, but male chauvinism was not suggested. In fact, Mrs. Hewlet, the first English woman to earn an aviator's license said, "Women will never be as successful in aviation as men. They have not the right kind of nerve."

The idea that flying was a young man's game was born early. "Below 20, boys are too rash; above 25, they are too prudent," declared a pilot in 1911. "The youth under 18 has shown itself to be bold, but he lacks judgement, and men over 26 are too cautious," said ace Raoul Lufbery in 1916 in an age-span guess for fighter flying which is reasonable today.

"Colonel" Samuel Franklin Cody, the illiterate Texan who became the first man to fly anywhere in the British Empire, confounded all attempts at explanation, having begun his astonishing career at age 47.

"The successful pilot must have a quick eye and steady nerve," ventured W. J. Abbot in 1918. Apparently convinced that the nerve idea had been overstressed, "Avion" (whoever he was) reassured young readers in 1919 with, "The pupil need not feel nervous. Only one is killed for every 120,000 miles flown." He provided no statistics for injuries, however.

Reading between the lines, it becomes obvious that the typical reporter was simply relating, with considerable embellishment, what he had witnessed from behind the airdrome fence. He was entranced but desired no actual involvement. Much of his observation was based upon guess and hearsay. Which is not an unfair description of today's typical newspaperman writing about aviation topics. In most cases he doesn't have a clue.

Collins, who began his book with the thesis that almost anyone could fly, so modified his view as he rambled on that he contradicted himself: "The aviator must be born, not made; he must cheerfully face constant danger; he must enjoy more than average health." He put stock in a number of "highly ingenious scientific instruments" devised by medical experts to test candidates for military flying — the "chronoscope and pneumograph" being employed to determine the "curves of an examinee's personal equation." As none of the medics had ever been up, one wonders what they were looking for.

Once airborne, keeping cool was the ticket as Beaumont saw things. "Presence of mind is invaluable when the flying man is in a fix through unexpected events, and these are numerous. In fact, is not aviation a mere succession of unforeseen incidents? The easiest aerial trips are accompanied by the most extraordinary surprises." On occasion they still are.

"Flying has a curious effect on a man; the strains on the nerves are enormous," observed Edgar Middleton in 1919. "The physical, as well as moral, strain is far greater than commonly supposed," said Collins in another book. "The air man must possess absolutely untroubled nerves." Ringing ears, frostbite, nausea, and fatigue were listed as making flying a miserable experience more often than not. "Many air men dread to sneeze at a critical moment," he went on. "A number of bad accidents have been attributed to a sudden sneezing fit which caused momentary loss of control." It is still unwise to fly with a bad cold.

High altitude was regarded as a special peril. "The brain is frequently so affected," said a reporter in 1915, "that aviators have been known to go sound asleep on their machines. Only the sturdiest bodies can stand the strain of long flights."

No man could long endure the rigors of it, said Middleton. "The strain rapidly puts an end to the 'flying life' of the average pilot," the flying life being defined as, "the period during which a man can fly a machine in the air with the same ease and comfort he would experience in the safety of *terra firma*. Suddenly will come a day when he is so filled with terror at the mere thought of flying, that he dare not attempt any of those little tricks and stunts upon which he formerly prided himself. After a long rest, however, a pilot can usually fly again as well as ever."

Technically, a great distance has been covered since that quaint and perilous era, yet the human factor remains as much a mystery, intensive physical examinations and psychological tests notwithstanding. Motivation remains immeasurable; the characteristics required of a good pilot only reveal themselves in practice and may surprise a student more than his instructors. One aeronaut nailed it down way back in 1911 when he said, "It is impossible to decide in advance whether a pupil will make a good air pilot."

That's still true.

4.

Three-Chapter Book

Airline history can be recorded in three broad chapters: the pioneering era which began with post-World War I airmail experiments and lasted through the Douglas DC-3, the advanced propeller days which ended with the Vickers *Viscount* and Lockheed *Electra* turboprops, and the jet age which was ushered in by Britian's deHavilland *Comet*. It was my good fortune to watch the story unfold almost from its beginning and to gain an insider's glimpse of each stage. All of the ingredients of great drama are in it — tragedy and humor, failure and triumph.

The most interesting chapter? The first. We vets who hired on in the late 1940s got in on the tag end of it. *Constellations* and DC-6s were taking over premium runs but DC-3s still moved most of the loads, operating primarily along airways laid down before the war. The busiest terminals boasted an Instrument Landing System each and the first omniranges were operating, but most navigation on DC-3 routes was accomplished with low-frequency ranges, homing beacons, and fan markers, and we worked the control towers at fields that had them on 278 or 396 kilocycles when close in.

There was no direct contact with Airway Traffic Control; clearances were telephoned to company operators who relayed them to us via high-frequency radio. Pilots had to carry thick bundles of sectional maps, to memorize all landmarks along their routes, and to prove once a year their ability to copy Morse code at six words a minute. Two groundschools, two checkrides, and twelve hours in the Link trainer were required of captains annually. Copilots did the Link but got no school or checks, it being assumed that their skippers, who wielded the power of job life and death, kept them on their toes. All training and check rides were conducted in the airplane itself.

This was the order of business established when the first two-pilot airliners were introduced a decade earlier. Many of our seniors had been there at the birth of the industry. They could tell you about the Ford and Boeing 80A trimotors, the single-engine Hamilton and

Lockheed *Vega*, and even the deHavilland DH-4 with a Liberty engine — and would at the slightest show of interest. The short life span of our industry is illustrated by the fact that many of them are still alive.

Virtually all chapter one air travel was completed within the first 10,000 feet above sea level, the only reasons for going above 6,000 being to clear terrain, avoid ice, or get a smoother and/or cooler ride. United shuttled north and south on the Reno range, struggling up to 14,000 feet, the minimum enroute altitude for the westbound hop across the Sierra Nevadas. Eastbounds from California lazily spiraled down to keep the descent rate at 300 feet a minute for passenger comfort. Oxygen was available for the crew and any fares feeling lightheaded.

Whenever possible, flights were made under Contact Flight Rules. We'd go for weeks without bothering the CAA for more than authority to penetrate a cloud deck, canceling the clearance when back in the clear. In the hot southwest where we on Braniff worked, it helped to bring up the headsets and turn off the stack of radios which radiated heat like ovens. Navigation-wise, we were pretty much on our own in those pre-radar days and often took mild liberties with prescribed routings in order to inspect points of interest. From down low you could really see things.

One of my captains invariably left the airway approaching Waco, Texas, to fly near a town named West. Then he would call Waco Tower and announce us as, ". . . southbound Flight 23, northeast of West," and the copilot who did not find this hugely amusing even after hearing it a dozen times was in for a long month. Another operated through Amarillo without speaking to the tower at all. A single click of his mike button always brought immediate response. "Good morning, Flight 16, you are cleared straight in," and we'd land on the grass at the south end of the field. After firing up, a single click drew taxiing and takeoff clearances, all such advisories being acknowledged with two clicks.

Contact night flying in good weather was delightful. The old Three padded along with minimum attention and the familiar patterns of lights below were duly noted as they slowly rolled toward us and disappeared beneath the nose. On a clear evening several rotating beacons led away toward the horizon, marking the airway. The light line. These 36-inch searchlights, rated at a million and a quarter candlepower, flashed 6 times a minute. They were spaced roughly 12 miles apart and located as closely as possible to the airway centerline. In addition there were two fixed course lights visible only if you were aligned with the route. Green flashes indicated an adjacent emergency field, usually a large pasture kept mowed by a local farmer, red that there was none. Each facility identified itself by flashing a number from 1 to 10 in Morse code. You had to know position within 100 miles for the identifications to make sense.

By 1950 these visual aids seemed antiquated and soon thereafter the network was dismantled. One airway beacon is displayed in the National Air and Space Museum. Above it in perpetual flight are a Northrop *Alpha* and Pitcairn *Mailwing,* the pilots of which found the light line a godsend.

We flew the airways as though they were highways, keeping to the right. Only during an instrument approach was it legal to be squarely "on the beam." "I want to see the beacons through my side window," reminded captains after dark.

Each old-timer kept a notebook of near-airport landmarks carefully sketched during fair weather arrivals. "Jeppesen's manuals help me find the airport but I find the runway," said one, and his circling approach in a snowstorm or heavy rain was something to watch — a minute on one heading, forty seconds on another, standard rate turn to a third, roll out and count to ten ("thousand and one, thousand and two, thousand and three . . ."), gear down, descend 100 feet, half flaps, watch for a water tank on your side, full flaps, throttles back — and the runway would swim into view right under the nose. I later watched the same men shoehorn big DC-7Cs into Chicago's Midway Airport with the same techniques and eventually employed them as my own, as when circling in a 707 to land north at old Kansas City Municipal. Eventually it became a lost art which was just as well.

Go back to all that? Not on your life! Nostalgia is for hangar talk and books. No one who ever rode through a Kansas line squall in a DC-3 or tried to keep it on an ice-coated runway in a gusty crosswind longs for any of it. Despite the remembered moments of humor and tranquility, the life was for the most part primitive and wearying. Occasionally it was absolutely hair-raising.

But I was glad to get a taste of the first days, and I value my working association with pilots who were there in the beginning. It brought the picture into focus and made me thankful for every improvement that came later.

5.

Paying the Price

Everything has its price. Like most would-be pilots, I was surprised to find the price of being taught to fly an airplane was more of what I had just escaped: school books, blackboards, memory work, figuring, the whole detested business. For every golden hour aloft we spent three in class and cramming for tests; that was the price and it had to be paid. You could be washed out in classroom before you saw a cockpit. It eventually dawned that groundschool had in effect begun a dozen years earlier. The sole reason I was able to keep my head above water in military schoolrooms was the learning habit acquired during formal education. Only the topics were different.

From all reports, the standards of public education are not what my generation remembers. Thirteen percent of today's high school graduates are said to be functionally illiterate, having failed to master the basics in grade school. A recent test given to 600 grade school children here and in Australia, Canada, England, France, Japan, Sweden, and Switzerland saw the Americans run fourth in geography, sixth in science, last in math. Though the Americans ranked in the top 35 percent nationally, the Japanese did twice as well in math. One-fifth could not find the United States on a world map. The young man or woman bent on a flying career should take notice and be aware that there has been no drop in the price that must be paid for any chance at a piloting career. In fact, the price has gone up.

Flying — that is, the piloting of aircraft — does not require great physical strength and but an average degree of coordination and dexterity. The mechanics of it are fairly easy. Flying is largely a mental exercise performed on an individual basis. The three crewmen in a jumbo's cockpit remain individualists, each with separate duties smoothly blended into the over-all effort. The single-engine pilot is a loner no matter what information he receives by radio. Those who picture a pilot as just another cog in the big machine are the first to point at him when things go wrong. He's the logical suspect, not them. The thrust of an accident investigation too often is based on proving that the pilot goofed and that there is nothing wrong with the system. So much for teamwork as applies to this trade.

A pilot must quickly absorb vast amounts of data before he gets a shot at the cockpit. It is no exaggeration that as much new information is covered during three weeks of airline or service preflight training as a college student confronts during an entire semester. Absorbing it quickly is not a knack picked up on the job.

Few laymen appreciate the book work behind every professional pilot or that he must continue to study all of his flying life. Study means concentration and self-discipline. It is hard work. There is little fun in sweating over books until you have a concept down cold, in learning to organize time and focus attention. There's no glamor in it, no coach to urge you on, no 50,000 fans to applaud success, no sweater letters, no trophies. There is little more than the satisfaction of independently accomplishing what could and should be done.

The demands and rewards of serious piloting are much the same. A good pilot strives to do a job that attracts no attention beyond his own ranks (Bob Hoover and the Blue Angels excepted). He forces his attention to a wealth of details which have little bearing on his work, never forgetting that complacency is lethal. His satisfaction comes from meeting the challenge, solving the riddles, and setting brakes without incident. His lifelong aim is to avoid outside attention, to stay out of the news.

If someone had sat me down at age 12 and spelled out the facts of flying, I would have been a better student. I would have understood that it was not what I was learning, but how, that mattered, that the ability to master a topic as boring as chemistry could be applied to a topic as tedious as aircraft performance. It would have made it much easier in the endless schooling that proved to be the price of my flying so many nice airplanes. Fortunately, a reasonable level of achievement was demanded in our early schooling; anything less was simply not tolerated. We did not emerge as candidates for Rhodes scholarships, but we had been taught how to study, how to learn, how to organize and file information. Most of us met the requirements.

The world of American education is not going to the dogs. Responsible parents and students alarmed by developments have demanded and are beginning to get better training in the basics and less time spent on extracurricular activities. Tough private schools by the hundreds have been established in recent years, and there are increasing numbers of no-nonsense public schools whose graduates are well qualified for university training — and for the good life that lies beyond. The pendulum has begun to swing the other way.

And it will swing back in pilot employment, no matter how bleak the picture now. The slack produced in military and civil flying by the defense situation, economics, deregulation, and other factors has largely evaporated. Today's student with the itch to fly can look forward to a cockpit career as interesting and rewarding as anything in the past — if he enters the game with an open and receptive mind well trained to learn.

"If only I could go back and do it again" — what sadder reflection is there? In my case, this: that I didn't appreciate what my parents and teachers did for me, rather than to me, by expecting my best. They did not always get it, but their old-fashioned ways, which are now being given a second look, produced enough results to make success possible. I did one thing right. I wrote to Mr. Davis, our high school principal, to acknowledge the debt I could never repay. He replied that he had received many similiar comments from grads. And, typically, he sternly reminded me that a diploma was but one rung of a tall ladder and admonished me to continue learning all of my life. The services and the airline saw to that.

The price? How modest it was. I was repaid a hundred fold for the energy I begrudgingly spent on learning during those first dozen years. The price remains a true bargain.

A recent survey disclosed that American kids between 12 and 17 spend more than 3 hours a day, 7 days a week, watching TV. When do they do homework? Following a schoolbus the other day I was enlightened. A score of teenagers got off at one stop. Not one carried any books.

6.

Nuts and Bolts

It is a popular misconception that every pilot worth his wings is a born mechanic. He could, if he had to, roll up his sleeves and restore a rough engine to sewing machine smoothness or resurrect a dead radio, given ten minutes with a soldering iron.

Once upon a time this was true. Orville and Wilbur were mechanics before they were pilots. Every short hop the early aeronaut coaxed from his contraption was followed by hours with wire cutters and glue pot. He had to be a mechanic to fly at all, and a good one to survive. The flier/technician image lasted into World War I, intensive class work insuring that a pilot understood his mount as thoroughly as the fitters who kept it in repair — in theory, at least. Mechanical-mindedness was seen as a prerequisite. If you had it to begin with, fine; otherwise it was drilled into you by mechanics.

In the field, most World War I pilots left maintenance to those trained for it, though some top aces spent hours in the canvas hangars. Frank Luke, the legendary Arizonan who claimed 21 kills in 17 days, worked on his Vickers machine-guns incessantly, examining each round as it was belted. Albert Ball, the boyish RFC ace who shot down 44 of the enemy, redesigned the windscreen of his SE-5A, mounted the spinner from a victim's prop, and increased his top speed. They were among the exceptions. The average pilot could credit survival to practice on the firing range and listening to his flight leaders rather than to technical savvy.

After the Armistice, it was back to the beginning for the few who stayed in the game. The barnstormer did his own repairs. Mail pilots were expected to solve minor problems out on the line; familiarity with the Liberty engines saved the life of more than one forced down in winter. A toolbox was standard equipment aboard early airliners; in fact, some lines would not hire a pilot without a mechanic's license. The Douglas DC-3 revolutionalized the industry but not cockpit training, which was intensified to the point that students had to memorize such irrelevancies as the gauges of Dural used in the airframe. This philosophy prevailed through the DC-6 and DC-7 and into the turboprop Lockheed *Electra,* the schooling on which was a nightmare.

The complexity of modern aircraft has led to a more operational approach in most quarters. When oil pressure drops, you've got a problem. Knowing who built the pump, how it works, and where it's located does not supply the solution. A pilot only needs to know what he can *do* about a problem from where he sits in the cockpit. Not all airlines subscribe, even today; some, for example, still require students to draw from memory the entire electrical and hydraulic systems, a considerable feat which proves little.

Thus a pilot continues to be regarded by many as primarily a technician in charge of a complicated machine, rather than as what he actually is — the manager of a complex exercise whose success depends upon far more than mechanical health. Such topics as traffic procedures, weather, airport facilities, and crew coordination receive minimum attention in many airline schools. Because few accidents are charged to equipment failure, the continuing emphasis on the machine rather than on its sensible operation is puzzling. And frustrating to all who suffer through it in classroom and flight simulator. As one fed-up old-timer put it, "I want to learn to fly it, not build one!" The traditional graduate's comment remains, "Now that all that crap is out of the way, I can go out and teach myself to fly it."

It was my good fortune to fly with many first-rate pilots yet no more than 1 in 50 showed a keen interest in the airplane as a machine. Most endured the nuts-and-bolts lectures, memorized enough of the numbers to pass the writtens, then promptly forgot most of them. They were all right there in the book if you needed one. There is little apparent correlation between technical aptitude and piloting skill. The pilot with a real interest in mechanics is to be envied, during groundschool at least. He laps it up, revels in learning exactly how each component performs, and spends time in the hangar on the fine points. He enjoys the course and breezes through the final, but none of it necessarily makes him more competent than his classmates. Indeed, a pilot can become so obsessed with mechanical details as to lose sight of the over-all picture, and that's dangerous. I've seen it happen.

I remember many 25,000-hour veterans of exceptional skill who couldn't wire a door bell. What they lacked in mechanical ability was more than offset by what the British call "air sense." We'd say they had the big picture.

Every pilot has to solve the puzzle for himself. Some of the pieces are provided in school but most come from personal experience, observation, and self-imposed study. He recognizes that much of what he is force fed in the process of earning and maintaining licenses is of meager value and that safe flight is based on common sense, not published procedures. Flight manuals are written by technicians to please federal regulators; planes are flown by pilots. The three specialties argue more often than agree on how best to get the job done. Disagreements usually are settled by those who regulate training and write the rules — non-flying company people and the feds.

Much training remains narrow and biased for several reasons, one being inertia. Many an inane practice is continued simply because "it's always been done this way." Resistance to change is one cause of today's chaos. And there's the legal aspect. A jury hearing a crash damage case, after thumbing through the handbooks issued to a luckless pilot and seeing the tests he passed, is certain to be awed and conclude that his training cannot be faulted. That it may have dwelt more on theory than practice is difficult to demonstrate to laymen.

A typical training program and the manuals distributed are as interesting for what they omit as for what they include. For example, few of the manuals I've seen cover icing except in a general way. What is the proper glycol/water mix for pre-takeoff hosing down of wings and tail? How can a pilot insure that his aircraft is receiving the right mixture? How long is it effective in various temperatures? How much of the white stuff that accumulates between the gate and the runway is going to blow clear during the takeoff run? The manuals deal in generalities, not specifics. The ultimate go, no-go decision is pointedly left to "pilot judgment." None of the ground crowd wants to share the limb he sits on.

I do remember, 35 years later, that the DC-6's tail skid strut pressure was 700 pounds per square inch, plus-or-minus 40 pounds. There was no way for a flight engineer to check it during his walkaround inspection, nor could he do anything about it had he known the pressure to be out of limits. But we had to know the numbers to pass the tests.

7.

The International Riddle

We boys reared during the Great Depression believed that any airplane worth talking about had "Made in USA" stamped on it and a red-blooded American on the stick. This was the definite impression given by air writers of the day, Hollywood scenarists included. We swallowed it whole, as did most of our elders.

One thing did raise questions in my mind. An aunt in England sent books at Christmas that told another story. British authors, after conceding that the Wrights most likely did invent the "aeroplane," strongly suggested that subsequent progress had mainly been on their side. They reminded readers that no American design saw combat in the Great War and inferred we had not come far since. They listed speed records held by Europeans and airline services established years before ours. Alan Cobham and the Jimmy Mollisons of long-distance fame, bearded Capt. O. P. Jones, dean of Imperial Airways skippers, and Charles Kingsford-Smith were Empire heroes scarcely mentioned in our books. It made me wonder.

The Japanese received scant attention on either side, being dismissed as imitators. Indeed, if their copies of aircraft were as poor as those of toys, they rated little space in serious reporting.

The success of German air power in the Spanish Civil War startled our military experts, not that their reports received much press attention or influenced an isolationist Congress to vote more money for planes. Charles Lindbergh toured Luftwaffe bases and concluded that the Royal Air Force would stand poor chances in a war, but his dire warnings, like Billy Mitchell's, were shrugged off. Once war began in Europe, reports sent back by American observers prompted official action, but popular confidence in American technological superiority and flying skill persisted until that Sunday at Hawaii.

My own reeducation began 10 months before Pearl Harbor when I enlisted in the Royal Canadian Air Force. First I learned that nationality had nothing whatsoever to do with flying aptitude. Our two flying schools trained students from Canada, Britain, Australia, and

New Zealand as well as us mavericks from below the border. If anything, the Canadians were the best of the lot — that is, of 100 applicants from each nation, Canada produced the most sharp graduates. A study of the top ranking aces in World Wars I and II tends to bear this out.

In the Middle East I met Rhodesians and South Africans. They could fly, believe me. So could the French, Dutch, Danes, Czechs, and Norwegians. Poles with unpronounceable names who had fled to England picked up fighters assembled in West Africa and ferried them to Libya, no mean feat for men in their 60s, which some were. Watching them in *Hurricanes* and P-40s set to rest the notion that Americans had any edge. The reason we were all there to begin with was that the Germans knew something about building and flying airplanes. This they repeatedly and forcefully demonstrated.

American-built aircraft dominated RAF operations there. Most types performed well enough, but so did most made in Britain. A *Made in USA* builder's plate did not invariably mean the best. Some of the equipment shipped from here should have been left here. And, some British examples were as unremarkable.

I would not trade that experience for anything. It led me to wonder if the airplane which had so rapidly evolved into an awesome weapon might not after the war become a tool of peace (assuming that we won; we were not nearly as sure of that as the folks back home, judging from their letters). It was not a frequent topic, only an idea mentioned now and then. We talked about international air travel, never dreaming the airplane would so quickly ruin Pullman, much less Cunard. It was one thing to read about Germans or Australians or Greeks, to study their pictures and hear them on the radio. Suppose we could go to where they live in hours instead of days and at reasonable expense, meet them, and have them visit us — what then? Would we be better off for it?

This line of thinking naturally sprang from the congenial relationship we enjoyed. To be sure, we differed on many points, and argued, but none of us returned home with the itch to scrap with any of the rest. We got along well enough, so wouldn't our peoples, given the chance? Of course, being geographically close had not averted war between the same European neighbors twice in 20 years. I suppose we American and colonial outsiders thought that because we had to help out in war we should be more closely involved in maintaining the peace.

And we were witnesses to the steady stream of military aircraft shuttling back and forth across the Atlantic. There was no longer any question about the feasibility of trans-ocean air travel.

But those youthful views were mostly idealistic nonsense and it is almost embarrassing to recall them even at this distance. It was naive to conceive such a noble role for aviation. Or was it? It can be argued that travel is narrowing rather than broadening, that it tends to cement preconceptions. Such can be the effect on a Londoner who sizes us up after a week in Chicago or an American who thinks he understands Britain after a week in London.

Some travelers are apparently so bent on not liking where they're going one wonders why they go at all. They come home with tales of bad plumbing, rude taxi drivers, and tourist traps, their worst fears confirmed. It's never again for them. Others gush. They bore friends with snapshots and begin saving for the next time. So what's the bottom line? Is bringing people together accomplishing anything? After listening to a great many international passengers, I don't have the answer.

But I am optimistic. I say it is working, or at least, it is helping. The knowledge that at this very minute thousands of people are speeding to distant places they have never seen, taking trips made possible by modern aviation, is encouraging. Back home they read of happenings datelined Paris or San Francisco or Manila with special interest and understanding. They were there. They should also be more aware of themselves, how they fit into the global picture and what they have — or miss — in the part they occupy.

Has this been a step in the right direction? I like to think it has.

8.

Where We Worked

The forward thinkers have had no luck at all with "cockpit" despite their success at renaming just about everything else. Years ago they changed aeroplane to *airplane* — at least, on this side of the Atlantic. I still like aeroplane. They changed throttles to *power levers* when the turboprop Lockheed L-188 arrived, then to *thrust levers* for the 707. More descriptive, they said. Mixture controls became *start levers* and Kollsman was modernized into *altimeter setting*. Ask today's tower man for the Kollsman and he'll come back, "Say again?"

Chief Pilot became *Manager of Flight*, Dispatch was changed to *Operational Control Center* (*"OCC"* was acceptable as the telephone response, however), and Ticket Agent to *Passenger Service Representative*. Excessively stupid vice presidents were no longer canned; they left *to pursue long-term career goals*. Same thing.

But they got nowhere with "cockpit," not that they didn't try. On the new L-188 *Electra* we were told to call our working quarters the *flight deck*. We laughed. On the 707 they would be known as the *control cabin*. More laughter. "Cockpit" it remained everywhere but in the equipment manuals. I cannot speak for the Air Force; any outfit that refers to its airplanes as *inventory* probably has a new name for the place where the pilot sits.

I like "cockpit." (To be sure, if you ride the airlines long enough, you're going to hear some pompous captain come on the PA with, "This is Captain Holcombe on the flight deck. . . ." Every line has a few of him. He'd call it the "eagle's nest" and fly in a chicken suit if it was in the book.) Cockpit is a good old term that ties things together — like windsock. It brings to mind planes I could only read about — Spads, Fokkers, huge Gotha bombers. I'd sit for hours studying a picture of where Billy Bishop sat in his Nieuport Scout. And those I could only watch — P-12s, Ford trimotors, DC-2s — and those I came to fly. After all, a pilot thinks of a plane from his past in terms of where he sat, remembering the view, the feels, the sounds, the smells, and how he adjusted to the arrangement of controls and instruments.

Mention the P-51 *Mustang* and I'm right back in that narrow cluttered fuselage, peering down that lean nose through the propeller's shimmering arc, feeling the high-frequency vibrations throughout my body, the thunderous, deafening explosions of 12 cylinders streaking back past my ears from the Rolls Royce's short stacks. That was a cockpit to remember. I could find the starter switch today, 37 years later.

Mention DC-3 and I remember the rain capes, two oilcloth sleeves, and a butcher's apron which helped in heavy precip, though the pool forming in your lap always funneled into your shoes. The subsequent squishing walk through the terminal raised eyebrows. I have not forgotten the bone-jarring shudder that was the penalty for landing with the tailwheel unlocked, after which boner we'd avoid the terminal. The large Automatic Direction Finder was mounted face up behind the throttles. It was a standard joke to tell a green hostess it was the "Virgin Indicator" and tune in a station behind us. When the needle swung round to point at one pert redhead, she said, "You'd better squawk that one."

The DC-4 cockpit was so wide there was a set of throttles for each pilot. What the airlines billed as a DC-4 was not a DC-4 at all, but a reworked surplus Army C-54 with fuselage tanks behind the cockpit removed. The result was a tail-heaviness which tended to extend the nosewheel strut, locking it straight ahead for retraction. If too many fares sat in back, you had to ride the brakes, dipping the nose for turns. Taxiing was accompanied by a not unpleasant chorus of hydraulic squeals and groans as if someone downstairs was tuning a pipe organ. The -4 also leaked like a screen door and carried as standard equipment rain capes for both pilots.

Our pro-Douglas bosses stunned us by purchasing two tired old L-409 *Constellations*, the original short model. A few of us what-the-hell types signed up for training. "Good morning, sports," said the Chief Pilot when school began. The cockpit was like three in a phone booth compared with the DC-4 and -6, yet comfortable enough once you squirmed into place. Good seats. Your head was right next to the smallish windows so visibility was adequate by transport standards.

It was a habit in Douglas equipment to grasp the glare shield above the instrument panel while adjusting the seat. We found another way after one ride in the *Connie*. Out of sight up there were trays containing exposed red hot heating coils, crude devices which helped keep windshields clear of winter ice. They were dubbed "bread warmers." You grabbed a red hot bread warmer but once. The Wright engines were prone to smoke while idling, which occasionally prompted

Dallas tower controllers (our pair were the only *Connies* operating into Love Field then) to advise, "Braniff, you're smoking," to which one testy captain replied, "It's old enough to smoke, isn't it?"

The Convair 340/440 series had big cockpit windows, good seats, and a sensible arrangement of the things you needed to see and touch. Its copilot was the busiest fellow in aviation during taxi out — copying the clearance and watching for ground traffic while adjusting cowl flaps plus a dozen other things, at the same time slowly easing his side window closed so as not to pop passenger ears. One of my skippers always left the gate like A. J. Foyt coming out of the pits and headed for the runway at a mile a minute, all the while yelling, "Come on, Morgan, let's get this show on the road, damn it!" I got it all done by the time the gear retracted.

In the Convair you switched on the water injection coming over the fence in preparation for a possible go-around. This caused both engines to miss a beat, something we soon accepted as normal, but we never got used to the sudden awful silence that resulted when we neglected to turn it off after reducing to climb power. We'd level off, lean the mixtures, and in five minutes or so the little tank would run dry and both sides would expire. Technical doubletalk was used in a P.A. explanation, none of which made the least sense.

The DC-6 oil-canned, that is, its fuselage panels popped out of it because of changing cabin pressure. These ominous underfloor snaps and thuds were invariably heard as we fretted over a sick engine or in rough air. Equally unnerving was duct rumble in the first 707s caused by loose duct work, chattering air valves, or whatever. The clatter and booming down below suggested the main spar might be coming unnailed. "Duct rumble," we'd tell each other, praying it was, while we glared at the flight engineer until he fiddled with his pressurization switches and got it stopped. Someone had to pay for the fright.

The 707's pilot seats were not directly behind the control wheel columns, this odd offset never being explained, not that it mattered. You never noticed it when flying. I never got over thinking about the 707's engine pods slung down there beneath the wings, quite close to the ground. Touch down with more than six degrees of bank, they warned, as when countering a crosswind, and the outboard pod would contact the runway, so we learned to crab awkwardly down to the runway with wings level and nose slewed into the wind, wondering how much side load the landing gear could take. No wonder the main struts appeared to have been salvaged from steam locomotives.

The *Electra* had the best air conditioning system of any airliner before or since. Its 13-ton freon system was a godsend to crews who worked summer schedules in the blistering southwest. But those seats! The chair itself was not bad but was mounted too low; your legs went straight out so that tail bone supported torso. Half an hour in moderate chop would loosen your fillings. Many *Electra* pilots developed serious back problems. "I can tell what you're flying by the way you walked in," said my FAA physician.

We lost an *Electra* in Texas due to wing failure less than a week after it arrived from the factory. Then Northwest lost one in Indiana the same way and we were ordered to reduce cruising speed pending determination of the cause. Any unusual sounds in the cockpit during the few months captured a crew's complete attention. One hostess, in an ill-conceived attempt to make light of our dread, crept quietly into the cockpit at 2:00 a.m., picked up the metal-covered logbook and dropped it flat on the floor. The engineer and I nearly jumped out of our skins. The captain, as relaxed a pilot as I ever met, never moved a muscle. He slowly turned and stared at her. "Why don't you just get a gun and shoot me?" he said. He said more after she was gone.

The 747 cockpit had Boeing stamped all over it. It was a repeat of the 707 but with certain refinements and lots of new gadgets. The big difference was the outside view. You looked out over the terminal building and even across those mini-jumbos, the Lockheed 1011 and Douglas DC-10. You felt conspicuous and self-conscious way up there, though it was difficult to be seen except by the crew of another 747. I never think of that cockpit without remembering my first transition ride. While I struggled around the pattern making touch-and-gos, my companion trainee explored the main passenger cabin downstairs. He came into the cockpit and tapped me on the shoulder. "Do you have any idea of what's following you around back there?" When he flew I went down and walked the aisle from nose to rear galley. Ninety-six steps. That impressed me as much as the light touch required to change the unbelievable monster's direction.

9.

Gone Are the Ships

Gene Darcy was that clever California entrepreneur who bought up all the *Queen Mary's* furnishings when the great liner was retired and made into a tourist attraction at Long Beach. He sold them as souvenirs across the country. You could buy a dinner plate, ashtray, life jacket, or chair from a drawing room. He gambled on public interest in the famous ship and the gracious way of travel she represented, and he was on solid ground. When the *Elizabeth* was sold, he was first in line for her appointments. It was said he earned a million from his unique enterprise.

Darcy gambled that his love of ships was widely shared and, of course, it was, not only by the millions who sailed in them but by those who stood on shore and watched, awed by their massiveness, entranced by their regal bearing, their elegance, and their grace. Many the time on a New York layover that I walked to the Cunard pier at high noon to watch the *Mary* leave. And what a spectacle it was, the massive black hull, gleaming white superstructure, towering red funnels, all of it sliding slowly back into the Hudson with that booming, soul-stirring foghorn echoing through Manhattan's canyons. If that didn't make you long for faraway places, nothing could.

That grand departure never failed to remind me of a certain warm morning in 1942. Our convoy had weathered an Atlantic gale and eluded — all but two luckless freighters at least — a U-boat pack off the Portuguese coast. We were zigzagging southward toward the Equator when the *Mary* appeared hull down on the horizon, its silhouette unmistakable despite the flat gray camouflaging. Within a half hour it swept into full view, steaming fast and entirely without escort, exchanged brief semaphore signals with our leader, and then became but a dot ahead where sea and sky met, leaving only a trace of smoke to mark its track. Strictly against regulations, I opened the port-hole and took a picture, which I still have.

Years later an item in our paper about the demise of the old and famous P&O Line, from whose *Viceroy of India* I took that snapshot, prompted a letter to the editor about that incident. It was published. That evening the phone rang. The caller was a druggist who had read my letter. We compared dates. "I remember your convoy very well," he said. "I was aboard the *Queen Mary*." Small world.

Standing on the bridge of the *Mary* in its final anchorage at Long Beach I remembered a postwar day at New York. "I left Floyd Bennett Field in a P-51, turned homeward, and there it was moving outbound through the Narrows, right where the Verrazano Bridge crosses today. What a majestic sight. I circled once then crossed ahead at 500 feet from left to right (port to starboard?), rocking my wings in salute, wondering if it might respond with a short blast from its horn. But it paid no attention; there was no acknowledging wisp of steam from the front funnel.

At one of Darcy's sales I bought a 16-page document from the *Queen Elizabeth* labeled, "Daily Engine Log Book." In it were recorded a wide range of temperatures, times, speeds, pressures, and so on, entries being penciled in once during each four-hour watch. This particular log was kept in the forward engine room during Voyage No. 453 commencing on 13 July 1967. In general arrangement and purpose it was similar to engine logs my flight engineers kept aboard our jets.

Voyage 453 took the *Elizabeth* from Southampton across the Channel to Cherbourg, a run completed in 3 hours, 15 minutes, at reduced power (average revolutions per minute — 101). The crossing to New York was completed with a "steaming time" of 6,786 minutes which works out at 4 days, 17 hours, 23 minutes, during which time the propeller revolution counters noted 1,119,050 turns. Yes, I also thought it was *propellers* on aircraft and *screws* on ships but the revs column was headed, "Propeller Revolutions." The margin notes were interesting. On the return run, Voyage No. 453 East, this appeared on the 21 July sheet:

Fog Standby — 3:40 a.m.
Watertight doors closed — 3:41 a.m.
Full away — 5:00 a.m.
WTD open — 5:01 a.m.

Strangely, the great liner was operated on local time using the 12-hour clock. Westbound the sheets are marked, "Clock retarded 1 hr @ midnight." Eastbound, it was advanced. An average rpm was 165. One day's reading ranged from 145.2 to 148.1 (high seas?).

Whatever the reason for slowing, the following day showed a steady 171.7, possibly to regain lost time.

Power variations were once used by us to offset the effects of enroute winds, but the practice was abandoned during the fuel crisis. The new plan was then to milk maximum mileage from each pound of fuel and arrive when you got there. I showed the ship's log to one of my flight engineers. "That fellow stayed on top of the readings," he said, "but I keep a neater log," and he did.

Aircraft and ships are in many ways similar. The good and bad of life in the air are not unlike what has been known to sailors for centuries. The machines differ, not their purposes or the perils faced by those who man them. In some ways the Air Age is a repeat of the Sea Age, or a continuation of it.

Darcy liked to say that the Air Age has nothing to match the charm and comfort of sea travel. Although I spent my working life in the industry that throttled ship lines, it is arguable that our eight-hour dash to London is a poor substitute for those five fabulous days at sea. We sold speed and the chance to spend your time in England instead of enroute. That is where we had the ships and we beat them cold. But, admittedly, getting there is no longer half the fun. Getting there now is not enjoyed but endured.

The *United States*, pride of our merchant marine, the ship that knocked ten hours off the *Mary's* best transatlantic time, has for years been stored at Norfolk. Not long ago I sailed right past it in a small sailboat. It sits there, abandoned and faded, a depressing sight for anyone who remembers how things once were.

Gone are the great ships and gone the gracious way of life of their times. Gone for all practical purposes are the French Line, the United States Lines, and Cunard. Now it's Air France, Pan Am, and British Airways. The airlines own the Atlantic, the Pacific, and all the other oceans. Thinking back on the rise and fall of sea travel, one wonders where on the curve the airlines are today. So far there is no hint of a replacement for the jet airliner but then, the same was said of ships. And trains. And horse-drawn carriages.

It is rather sad that great ships and planes became locked in a death struggle. It was inevitable, but in our failure to find places for both in the scheme of things we lost something.

10.

"Odds 'n Ends"

If I had anything to do with designing an airplane, it would bear a model number which included "47." *Four seven*, in that order, is a winning combination. There was the Boeing 247D, the first all-metal, low-wing, multi-engine airliner; it lopped seven hours off coast-to-coast airline schedules and outran the Army's hottest fighters. And North American's 0-47, the most widely used prewar observation aircraft. The Boeing B-47, Douglas C-47, Republic P-47, and Boeing 747 are pure gold, of course, not to forget the Bell 47 helicopter, some 5,000 of which were built during 27 years of continuous production. I cannot think of a single flying lemon that bore the number, "47."

• • •

When we lived in rural Georgia during the Great Depression, my father regularly wrote to relatives in England. His letters went by train to New York, steamer to Southampton, train to London. When everything clicked, they were delivered in 7 to 10 days. Sometimes he had replies — way down there in Monticello, Jasper County — two weeks later. Thus, the average speed was 19 miles per hour each way, which was not very good. That was long before the 600-miles an hour jetliner or transatlantic airmail service of any sort.

Fifty years later I lived 26 miles from the Dallas-Ft. Worth Airport and corresponded with an aunt who lived, strangely enough, 26 miles from London Gatwick Airport. DFW and LGW were connected by daily nonstop 747 passenger and mail service which spanned the distance in 10 hours. The average my-home-to-hers time ran seven days, which says that half a century of transportation progress had brought no improvement whatsoever in average speed. And I could not even depend on that. One letter to England took 12 days, the reply 13, which translated into the average speed of the Pony Express.

My mailman blamed them, her postman us. "Look at it this way," joked my mailman. "We send your letter to England for a dime. The other 30 cents is for storage." He was quite sure they stored it over there. An inquiry of the Dallas postmaster resulted in a two-page letter of explanation which explained nothing. The question asked, "Since my letter only flies for 10 hours, what do you people do with it the rest of the time?" went unanswered.

Postscript: about that time the papers reported that the U.S. Postal Service was sending revised rate information to all its stations — by United Parcel Service.

• • •

Do we ride *on* airplanes or *in* them? Old books have pictures with such captions as, "Mr. A. V. Roe *on* his machine at Hendon," and he was indeed on it, there being no way to get into it. Today a pilot will say, "I'm checking out *on* the DC-9," which is impossible unless he's learning to wash it. It makes as much sense to say you're going to town *on* your Buick.

Rail and sea travel is no less confusing. The Cunard office windows on Fifth Avenue used to display pictures of movie stars who had just crossed *in* the *Queen Elizabeth*, though most passengers spoke of sailing *on* it. If astronauts go *up* (rather, *out – away?)* in their space capsules, shouldn't skydivers come down *under* their parachutes? And, how can you go up *in* a balloon? The FAA should clarify these issues once and for all. You can bet they're working on it, because they've come up *(out?)* with rules for everything else.

And what about directions? When we lived in Dallas, we flew *up* to Minneapolis, *down* to Houston, *out* to Denver, and *over* to Memphis. Then we went *back* home. No problem. But what about Los Angeles? (I suggest *way out* is appropriate for LA.) The IRS should tackle this one; they know where everyone is.

This problem is international. I was sent to a specialist in London for minor surgery. They said I'd find *Mister* Taylor (that's how they address their surgeons over there) at his offices *in* Harley Street but I found him *on* one side of the street, needless to say.

• • •

It was about 1950 that Howard Hughes demonstrated his new anti-collision device by flying a *Constellation* toward mountains. Bells rang, lights flashed, and he veered away with room to spare. Most impressive. Since then there have been a number of excited announcements about devices which hereafter will prevent collisions between planes and the ground — and with each other. Recently the FAA said it is bearing down on development of an anti-collision system that really will work. Don't hold your breath.

• • •

William Arthur Bishop wrote a biography of his famous father, Canadian World War I ace Billy Bishop. The point is made that Billy was a somewhat ham-fisted pilot who smashed up a number of planes in the process of knocking down 72 of the enemy. His landings were a squadron joke. Manfred von Richthofen, the vaunted 80-victory German ace, was also a less-than-spectacular pilot who nearly washed out during training. Bishop and the Red Baron fought each other to a draw more than once. France's frail, sickly Georges Guynemer survived more than 600 aerial combats and claimed 54 victories. Britain's Mick Mannock had to memorize the eye chart to enlist, yet he shot down 73 of the enemy. The more you read about the first war in the air, the more unbelievable it seems.

• • •

Have you noticed that when they send one of the enormously expensive shuttles into space they pick an old half-bald geezer to bring it back in one piece? When I was in the Army they put you behind a desk as soon as you began to understand what flying was all about. Obviously, times have changed. The graybeards who have been slicking on those monstrous gliders certainly haven't spent their years doing paperwork. Every time they step out, every retired pilot grins from ear to ear.

• • •

My hat's off to Larry Walters, despite his foolishness. The typical armchair pilot sits and talks, but not Larry. He's the fellow who some years ago tied 14 weather balloons to his lawn chair and soared up to 16,000 feet. Delta and TWA flew right past him. I'd like to hear those cockpit voice tapes. How did they report passing a man sitting in a chair three miles up?

Larry got cold and numb and called for help on his CB radio, but soon realized what some pilots never learn, that a radio is useless in most emergencies. So he punctured some of the balloons with a BB rifle and down he drifted to land without injury. The FAA took a dim view of Larry and charged him with operating "an unauthorized vehicle in a controlled air space." It was, of course, an extremely dangerous stunt, a hazard to aircraft for one thing; any repetition had to be firmly discouraged. But don't you agree that the National Air and Space Museum should have that lawn chair? There's not another one like it anywhere.

• • •

The Deregulation Act was passed in 1978. In the immediate aftermath, 15 carriers merged or folded and 50,000 employees lost their jobs, 5,200 of them pilots. Mr. Alfred Kahn, the economist whose efforts helped convince Congress that the world's finest airline system needed a complete overhauling, said the upheaval was painful, but necessary. He sounded like a man who never underwent hemorrhoid repair surgery comforting one who just did.

• • •

Watching the Dallas Cowboys perform, it is not difficult to believe that coach Tom Landry flew four-engine bombers during World War II. He was in B-17 *Flying Fortresses* out of England, they say. His cautious, conservative approach to every situation and the complexity of the plays he sends in do seem to reflect the philosophy of a pilot trained to doggedly press on according to plans laid down before takeoff. I sometimes wonder how the Cowboys would have fared all these years had Tom flown fighters in combat situations which dictated continuously changing tactics.

11.

Old Favorites

Of all the aircraft I have read about, seen or flown, the Curtiss P-6E remains a special favorite. Examine the sole survivor in the Air Force Museum at Dayton and try to visualize any improvement of its classic lines. There's no way;it's perfect as is, exactly as it was delivered to the Army Air Corps 50 years ago. Curtiss was a name to reckon with then. The 1933 *Jane's All the World's Aircraft* devoted seven pages to its products. Boeing rated three, Douglas and Lockheed two each, Grumman one. Curtiss also built remarkable engines, one being the 650-hp Conqueror that powered the P-6E. The sharp crack and snarl of that short stacked V-12 coming to life made my skin crawl. It would today.

Curtiss hung a pair of them in its big *Condor,* one of which was barnstormed by transatlantic hero Clarence Chamberlin. A hop in that ponderous fabric-covered biplane was the best four bucks I ever spent.

• • •

For all of its graceful shape, the P-6E was no more appealing to the eye than Boeing's P-12E and its Navy twin, the F4B-4. Those scrappy little beauties howling across the silver screen did more to fill the cockpits of World War II than all the recruiting posters printed. Similar in design and purpose was the Gloster *Gladiator*, an example of which is displayed in the wonderful RAF Museum at Hendon. A number flew from our desert bases in 1942. For that matter, the Hawker *Fury* was every bit the match of the P-6E, the throaty roar of its V-12 Kestrel no less soul-stirring, its curves no less inspired. I watched one fly and will never forget it.

The pre-World War II years of civil and military biplanes saw many interesting ideas, some downright beauties, but those four stand out for me. If there was a chance to fly just one, I'd want a week to think on it; then wonder if my choice was best.

• • •

Dr. Claude Dornier built a flying boat in the 1920s that intrigues me still. His Do.X had 12 more feet of wing than the largest 707 and a three-deck hull. It rose from Lake Constance on a test hop with 169 passengers and a crew of 9. All of this, remember, only 25 years after Orville flew 120 feet. Dornier's incredible giant was powered by a dozen Conquerors, six pulling, six pushing, mounted atop its wing. Imagine the glorious takeoff din. Though the Do.X plodded along at 108 miles an hour and had a service ceiling of but 1,600 feet, it completed a 35-leg tour from Germany to Africa, South America, the United States, and back home. Two mechanics 16 feet aft of the cockpit handled the throttles. I've often wondered about the pilot-engine room telephoned exchanges during approaches to land. This pioneering jumbo was destroyed during a World War II air raid.

• • •

Earlier in air history, the decade following Kitty Hawk was one of trial and error, ever exciting, often tragic. My favorite from that quaint era is Louis Bleriot's neat little cross-Channel machine. Its configuration — engine in front, tail behind, single wing, conventional landing gear, enclosed cockpit — pointed to the future though it was not generally adopted for 15 years. Although the first true fighter aircraft with forward-firing machine-guns were the Morane and Fokker E.IV, both of them monoplanes, distrust of single wings saw biplanes filling most World War I roles. Two of them appeal to me — the S.E.5A, a handsome and robust British design with a stupendous combat record, and the Sopwith *Triplane*, a curious creation whose almost dainty appearance belied a rugged constitution. Five Canadians flying "Tripes" scored 87 victories in 120 days with 2 losses. One of those shot down, Gerry Nash, was the genial officer of lofty rank who pinned on my wings 24 years later and confirmed that Tom Sopwith's unique design was in all ways a pilot's airplane.

• • •

Few high schoolers of the late 1930s logged more behind-the-fence time than this aficionado during which aching hours I grew familiar with the plane names of those days — Aeronca, Porterfield, Rearwin, Ryan, Spartan, Waco (pronounced "Wah-ko"), and even Stearman-Hammond, Timm, and Harlow. The Stinsons won the beauty contest hands down. I reacted to the SR-10C *Reliant* as I had to the first Model A Ford, "They've done it! It's absolutely perfect; no conceivable change could improve it." Whether the changes made since in either have been worth the price is arguable. Don't smile until you've driven a Model A and flown, or flown in, a gull-wing *Reliant*.

• • •

Fairchild made no attempt to match the Stinson's voluptuous curves, yet its Model 24 has nice lines all the same. The Warner radial engine edition was a delight to the eye and the *Ranger* in-line engine model was sharp indeed. As the one-time owner of a Fairchild 24 with *Ranger* power, I can confirm that it was, in flight, every good thing said of it. Another favorite of that period is the Buhl *Pup*, a pot-bellied little sport monoplane with open cockpit, external wing wires, and a three-cylinder, 45-hp Szekeley engine. I cannot explain why I'd walk past everything else for a closer look at a Buhl.

• • •

My love affair with the DC-3 stems not merely from its (then) immense size but for what it seemed to promise. Watching it vanish into the night with 24 souls on board, I could believe that flying was surely becoming a sensible and secure way to get there and that time would see more people riding airplanes and less aboard trains and ships. That would mean a growing industry in which I might find a place. Because it all worked out that way, I feel almost grateful to the Douglas creation that did more than any other aircraft to bring it all to pass. No airliner since exuded more grace in flight nor has there been one that, in its day, was more fun to pilot.

• • •

The C-46 is a favorite. Yes, it got off to a bad start and never lived down its unsavory reputation, but it remains a favorite and not simply because it was the first large aircraft I flew from the left seat. That old tub was built like a bridge; it could haul the load and go the distance in virtually any weather, but it did take some getting used to. Those who deride "Old Dumbo" never really knew it. Airplanes are like people — some are slow to grow on you.

• • •

I sat many a time in the cockpit of a Martin B-26 *Marauder,* my favorite among World War II bombers, itching to fly it, but the chance never came. That superb airplane also got off on the wrong foot, then overcame its bad name and ended the war with the lowest combat loss rate of any American bomber.

The P-51 *Mustang* would have been a favorite had I never flown it, but I did and it provided a vivid glimpse of the fighter pilot's unique world, an unforgettable experience for one whose career was tied to heavies. And there was the Hawker Mk. VI *Tempest,* the mere mention of which still makes my mouth dry. Imagine, if you can, strapping into a single-engine cockpit knowing you would scramble through 10,000 feet 2 minutes, 42 seconds after breaking ground. It was the size of our big P-47 *Thunderbolt*, but 4,000 pounds lighter and powered by a horizontal-H 24-cylinder Napier Sabre delivering 2,450 hp on takeoff. What more can I say? Only that it was an exceptionally handsome brute as well.

• • •

To paraphrase Will Rogers, I never met an airplane I didn't like. The transports in my log, from Ford trimotor to Boeing 747, left mostly favorable impressions. Of the propeller-driven types, the turboprop Lockheed *Electra* included, the Douglas DC-7C was the best of all. It had a perfectly delightful control feel thanks to its aerodynamic trim tab system and was rock solid on instruments. My enthusiasm was not universally shared, however, because of its engines, those awesome Wright R-3350 turbo-compound monsters with power recovery turbines.

Exhaust gases spun the PRTs (three per engine) at 19,000 revolutions per minute, if memory serves, recovering 135 hp each which was fed back into the crankshaft through fluid drives. The whole idea appeared hopelessly complex, yet the engine ran like a Singer — most of the time. A knowledgeable flight engineer was as essential to engine health as lubrication. With his analyzer he could monitor the firing of any one of the ship's 144 spark plugs and advise precautionary shutdown before things started coming unnailed. During climb, the PRTs ran white hot and the bright yellow flame of exhaust wrapped around the wing's leading edge enabled me to do cockpit paperwork without turning up the lights. What a piece of flying machinery that DC-7C was.

• • •

The Boeing 707-327C was the nicest of the jets. That was the big one, the long-range *Intercontinental*. I was fortunate enough to fly that magnificent airplane on the military airlift runs to Far Eastern points when every departure was at maximum gross weight and 10 hours the average time enroute. I can still hear those Pratt & Whitney JT3Ds as we rolled down most of 12,357-foot Runway 8L at Honolulu International, anxiously waiting for 175 knots to show on the indicators. Those turbines sang; I still have a cassette made during night departure at HNL. Next stop: Clark Air Base in the Philippines.

• • •

The 747? Huge, impressive in every respect, and fitted with wondrous mechanical and electronic aids to make work easier, more precise, and safer. It was a marvel from radome to tail lights. But I was never completely convinced that I was 100 percent in charge. It went where I steered, climbed and descended as I directed, and settled back to earth gently enough when I worked at it, but I never shook the feeling that I was akin to an elephant handler whose charge turns, not because his master prods him to, but because he agrees that it is time to change direction. Whereas you told the 707 what to do, you made suggestions to its big sister, or so it seemed to me.

• • •

My experience in non-transport aircraft is extremely limited, thus my favorites among them are based upon brief encounters or what I've seen of them. In the post-World War II period I greatly admired the Cessna 195 and, since, the 185. I can think of no other small personal craft that would please me more than a 185. As for small twins, give me an Aerostar 602P. I like its looks and performance figures. Among the small business jets, I'll take the Cessna *Citation II*. I flew one and found it near perfect.

A number of fighters of astonishing performance have appeared since my P-51 days, not one of which I did not admire. The Lockheed F-104 and Republic F-105 are my favorites, and the new Convair F-16 is breathtaking. I'd like to fly, or at least ride in, an aircraft that can accelerate to supersonic speed climbing straight up. Where can we go from that? What's left? Which is what I thought when they introduced the Model A.

Too many splendid aircraft have vanished entirely; only pictures remain. Oh, the shame of it. History-minded enthusiasts around the world are working hard to prevent the same fate for notable types of the recent past and we are in their debt. Many famed machines still fly, thanks to the Confederate Air Force and similar groups, while others are being restored for permanent display, which is not nearly as satisfying. I thought on this while admiring the P-6E at Dayton, its Conqueror forever stilled. Then I walked down the hall and was again reminded of the past by another silent relic.

Major Glenn Miller's trombone.

12.

The Human Factor

It was not often that we knew much more about our passengers than how many there were. We could figure load factor from that and load factor translated into red or black ink. Anything better than 55 percent meant the trip was profitable. Once in a while the number was translated into something starkly human.

We were flying a 727 schedule from the Dallas-Ft. Worth Airport to Los Angeles. Just before departure the senior cabin attendant came into the cockpit. "Captain, we've got a very ill child on board with a swollen face. A little girl about three. We'll keep an eye on her. I thought you'd want to know."

An allergy perhaps and she's being taken to a specialist? "Her mother says her insurance is good only in California." And she added, "No, I don't think it's an allergy." I didn't ask why she thought that.

We pushed back on time, lifted into a clear midday sky, and turned west. After we leveled at 35,000 feet, the copilot said, "I never heard of insurance good only in one state." That puzzled me, but in this age of mass air travel you hear strange things every day. I gave it no more thought.

Albuquerque lay over the nose when another hostess came up with coffee. How's the child now? She shook her head. "Not good. She can't stop crying. The coach girls have her in the rear galley and are trying orange juice." The mother was Hispanic and difficult to understand but it seemed the little girl also had ear problems.

Ear problems. I looked at the flight engineer's panel; our cabin altitude was 6,000 feet, slightly above the elevation of Denver, and it would have to be reduced to sea level prior to arrival at L.A. We liked to remain high until the last minute to conserve fuel, then drop down at 3,000 to 4,000 feet a minute at idle thrust. The engineer meanwhile brought the cabin down at 300 to 400 feet a minute, the resulting slow rise of pressure being unnoticeable to the average rider. But one with ear trouble, particularly a youngster, might have difficulty adjusting to even a slow rate of pressure change, in which case we would flatten our descent, traffic permitting. Unrelieved differential pressure in the ear can be excruciatingly painful. It's like a nail being driven in.

The engineer was already figuring. "If we start down 140 miles out I can keep cabin descent at 200 feet a minute or less." I agreed that we would ask Los Angeles Center for clearance to start down early. We continued cruising at Mach .79, our usual sedate fuel-saving pace. Groundspeed worked out at 440 knots which was given as 506 mph in a brief P.A. announcement about flight progress. Prescott, Arizona, fell behind and we set course for Twentynine Palms, California, the next check point. The copilot took the arrival procedure pages from his manual and laid them on the radar scope where we both could review them. The senior girl came into the cockpit.

After fifteen years of line flying an airline hostess could cope with almost any cabin situation on her own; she didn't bother the cockpit with petty problems. This veteran was clearly worried. "She is in extreme pain. It's a spasm one minute and listlessness the next. She needs a doctor badly."

We were then 115 miles from TNP, 250 from LAX. I tuned in Phoenix: 110 miles; we could divert and land there, saving minutes. I asked the engineer to increase power to maximum allowable cruise thrust, then go back and see what he thought. The copilot advised the LAX radar controller of our increasing speed and the reason for it.

"She is a little better," reported the engineer. "L.A. will be soon enough but she definitely needs medical attention." He added, "I feel sorry for her," which I did not comprehend just then. There was much to be done and little time to talk.

By telephone patch through our San Francisco office I gave the facts to the dispatcher in Dallas, requesting that a physician and ambulance meet us at the gate. A Spanish speaker would be useful as well. I repeated the facts to the LAX controller — very sick little girl, swollen features, convulsive, lethargic — and requested priority traffic handling and a straight-in approach to Runway 24L, the closest to our terminal gate. He came back immediately with clearance to descend at our discretion and to disregard all published speed restrictions including the 250 knot limit below 10,000 feet. Medics were awaiting us. He vectored five preceding trips out of our path, explaining to each that we were barreling down behind them with an on-board emergency. Little lady, I thought, whoever you are, and whatever your grief, you've got more good friends than you know.

We whistled down the final miles in shallow high-speed descent, landed, and hurried to the gate, the taxiways being cleared for us.

Several men waited in the passenger loader — two paramedics, two huge cops, a bearded young man with medical kit, company officials. Please remain seated until these people get to the rear of the aircraft, the passengers were asked, and all complied.

When the cabin was cleared, I went back. A hostess held the little girl while the doctor made his examination. I looked and was shocked. The child had been brutally beaten. Grotesque saffron bruises had almost closed her eyes, there was a cut lip and one ear was red and swollen. "There's more of it on her body," said one hostess.

You read about it, you hear about it, you know it goes on, but you don't believe it until you see its appalling evidence. It is horrible. You are overwhelmed by rage and frustration. What twisted thinking could provoke an adult to such monstrous cruelty?

The story slowly emerged. At first it was that the little girl "was like this when she woke up this morning." Then, under the doctor's patient questioning, the truth came out. A boyfriend was involved. The trip to California was a flight from horror. I was glad we had a hand in it.

Young children can become ill rapidly, and recover as quickly. The little girl sat up and took an interest in her surroundings. She looked at the girls who had shown her kindness, then smiled and waved. The bruises would soon enough be gone but what about the emotional damage, I wondered, wanting nothing more at that moment than to take her to the homes of sunshine and laughter where my grandkids lived, to be adored and spoiled as they were.

But I couldn't do that, so I joined the rest of the crew and rode to the hotel. Little was said on the way.

13.

Aches and Pains

The prestigious *New England Journal of Medicine* recently took notice of several work- and play-related infirmities heretofore unrecognized, at least, officially. Musher's knee and hooker's elbow have now joined housemaid's knee and tennis elbow. Musher's knee is a pain "caused by the sharp backward kicking of the leg by a dog team driver at the rear of a sled to spur his team to greater speed." Hooker's elbow afflicts ice fishermen, no matter what you thought. After days perched over a hole in the ice, "repeatedly jerking on a line attached to a stick," lateral epicondylitis can result. Not to mention a runny nose.

Also added to the list are dog walker's elbow, disco digit (fingers sore from snapping while dancing), and urban cowboy rhabdomyolsis (muscle damage you-know-where caused by too frantic a ride on a saloon mechanical bull).

We of the flying fraternity have our own unique aches and pains, among them:

Aviator's Arm: a noticeable skeletal deformity caused by carrying too many large suitcases through too many long terminal wings for too many years. Irreversible arm extension of up to 16 inches has been confirmed. One international captain was eventually able to scratch his knees while standing at attention, which he often did, just for the hell of it. Coincidental shortening of leg bones and sloping shoulders are common. Though painless, the condition poses tailoring problems in that the sleeves of an off-the-rack size 44 scarcely reach the elbows of an advanced case. A similar disfigurement has been reported by anvil salesmen with large sample cases.

Captain's Neck: a painful spasm produced when the head is tilted back, as when examining overhead panels through bifocals. In extreme instances, something gives, resulting in *Left Seat Lock,* the total inability to rotate the head back to its customary, eyes-forward posture. The pilot thus immobilized must complete the flight with bifocals worn upside down and walk off the plane with cap bill resting on chin. More than one copilot has been heard to say, sympathetically, "My captain has his head up and locked."

Coach Crouch: a passenger complaint incurred after sitting for hours in an economy fare seat. The afflictee appears to be searching for a dropped quarter. Fortunately, the problem responds to a few days in traction. An associated stress affecting riders not warned to wear loose fitting clothing on long trips is properly termed, *Coach Crotch.*

Second Officer Tic: an occupational malady induced by staring at gauges for extended periods. The symptom is rapid and uncontrollable blinking, each eye separately. Hallucinations follow. One flight engineer claimed that during a New York-Tokyo nonstop, the dials on his panel were transformed into merry little faces, some with goatees. He insisted that an oil pressure gauge winked at him and admitted that he winked back and tried to strike up a conversation. Doctors of psychiatry are seeking a treatment; meanwhile, many engineers go about with vacant stares, muttering and speaking to themselves.

Seniority Syndrome: a memory paralysis bordering on retardation common among airline captains in the over-50 age bracket. The ailment stems from preoccupation with job advancement. The typical sufferer can never recall a name, but can repeat without hesitation the retirement date of his 240 seniors. "Good morning, 23 March 1988," is his typical greeting. There is no known treatment or cure.

Latin Lurch: a disorder common among vacationers returning from a region reputed to have questionable drinking water. The ailment has previously been labeled with such non-medical terms as, "Aztec Revenge," "Memphis Murmur," "Egyptian Gut," "Green Apple Quick Step," and "Tourist Trot." An advanced case stumbles drunkenly along jumbo aisles, hands clutching lower abdomen, desperately seeking an unoccupied blue room. He has acute gastrointestinal distress which advances with lightning speed as altitude increases. The rider whose problem manifests itself in a small airplane has *Cessna Cramp.* The expression of utter hopelessness that is the trademark of "CC" is caused by the terrible knowledge that there *is* no blue room.

Subsonic Stomach: a strong suspicion that the hamburger hurriedly eaten prior to boarding is still down there in its original unchewed state — soggy lettuce hanging out, mayonnaise oozing, bun halves impaled on toothpick, pickle and little paper cup of gray slaw nestled alongside — and the whole greasy mess has solidified like cement. This is not illusionary. Digestion ceases during takeoff and, in fact, can reverse itself once there is no turning back. Many passengers have arrived minus the nourishment consumed before departure.

Checkride Skin: clamminess caused by excessive perspiring in an air conditioned flight simulator, usually noticed during the third attempt to complete a steep turn without wandering off altitude. Not to be confused with *Fed Fever,* the rapid elevation of body temperature occasioned by sudden and complete memory loss during an FAA oral exam. One examinee could not remember his wife's name, much less the stall speeds of his DC-10.

Thunderstorm Foot: loss of control of one lower limb when a large cell appears dead ahead on the radar screen. The subsequent sudden hard ruddering is explained to passengers as an autopilot malfunction.

Morgan's Mania: (well, did medical breakthrough end with Hansen and Hodgkins?) an incomprehensible state of mind which deludes airmen into believing they are the most fortunate of humans. Despite unwelcome and unnecessary interference by desk-flying experts who lack any real understanding of the activity they pretend to regulate, these strange individuals remain convinced they play the greatest game on earth. Possibly it is because they remain individuals in a society which long since stuffed most people into their assigned slots. Prognosis: contagious, incurable.

14.

Check Ride

We all hated check rides. All of us normal people, that is. I did hear one fellow say he welcomed his semiannual proficiency checks as opportunities to demonstrate his skill. Imagine flying a month with him. The majority detested all forms of training — groundschool, writtens, orals, simulator — and recurrent checks most of all. Recurrent training because it forced us to prove we could do what we had been doing all along.

But it was part of the life, as unavoidable as storms and sick engines, a price to be paid. The more rapidly a pilot advanced during periods of expansion, the more schooling required to acquire and maintain the necessary qualifications. He had to submit but he didn't have to like it.

Individual pilots approached the ordeal in different ways, a few simply by avoiding a new type of airplane, remaining with something familiar until it was phased out from under them. The horror stories — with little foundation — that attended the introduction of jets prompted more than one 25,000-hour veteran to take early retirement rather than try to comprehend the fearsome new machine. Others worked themselves into a lather before taking the plunge and took a month's sick leave when it was over. They were, in fact, exhausted and ill when it was over. Most gritted their teeth and got on with it.

There were right and wrong ways to tackle schooling. Where a good pilot had to continuously keep the over-all picture in mind and never become distracted by details, school required a reversal of thinking. The trick was to divide the program into its several stages, attacking the details of each while giving no thought to what came next. This was completely out of character for the pro.

First, the classroom work. You had to learn the new type from nose to tail and become intimately acquainted with its systems, limits, weights, pressures, speeds, and all the rest which meant memorizing columns of specifics rarely needed in practice, numbers right there in the manual should you need one. But federal inspectors did not allow open book written tests or orals. The curriculum was approved by the FAA, of course, and it was not unusual to have a fed sit in on an entire course, not only to verify that it was properly conducted, but to earn his own ticket. He sweated along with the rest of us.

Feds were not all of the same mind. You could draw a numbers man for the oral; if you had all the speeds and limits and pressures down cold you were in. Or an electrical or hydraulics wizard. Or an essayist like the one I drew for the 707 oral. "Tell me all you know about the anti-skid brakes," he'd say, injecting questions as I stumbled through what I remembered. The wise candidate answered each question — *period*. If the answer was, "55 amps," he gave that without elaboration. It was tempting to show the man how much you knew or thought you knew, but sooner or later you'd fly up a blind canyon if you persisted. The typical fed was thorough but fair, and he knew the airplane in question well enough to determine if you did.

With groundschool and oral out of the way, the worst was behind and it was on to flying. Originally, all flight training was conducted in the airplane itself, then we got half in the flight simulator and half in the air. Later, sophisticated new simulators with realistic visual airport displays on a screen before the windshields made it possible to conduct all training indoors. It is today entirely possible that the captain of a 420-seat Boeing 747 leaving New York for Paris has never before set foot inside the jumbo. Occupying his copilot's seat, however, will be an experienced check airman to oversee his performance during the first 25 to 50 hours of flight and through a number of arrivals.

The average student started off like a house afire, slipped rather badly the second day, and couldn't do anything right the third — or so it seemed to him. There were always a few who were good the first time at the controls and who steadily improved; it was discouraging to be paired for cockpit training with such an ace. Things usually picked up the fourth period and were likely to go well the fifth. No matter how a student assessed his own progress, his instructor probably noticed satisfactory progress from the first.

Cockpit training proceeded more rapidly and pleasantly for the student who had all checklists and procedures down cold beforehand. You simply could not concentrate on smooth handling while racking memory for a specific speed or flap setting. Every instructor had his own pet techniques and they might be at odds with a student's usual way of doing business. It was best to do it the instructor's way; he knew what the FAA wanted. It was a waste of time to argue fine points with an instructor, much less a fed. Out on the line you settled back into your own routine, adapting old tricks to the new machine.

At the end came the rating ride with instructor playing copilot and an FAA air carrier inspector watching. An Airline Transport Pilot license is valid only in the specific types of aircraft on which its holder was rated. The DC-9 captain moving up to the DC-10 underwent complete training on the new plane — groundschool, simulator, the works. And if a year later he was bumped back to the -9, he underwent an intensive refresher course and flight check prior to resuming his old seat.

All of us were afflicted with "checkitis" to some degree, a few wretched souls to the point of physical distress. A friend, upon seeing my name on the list to qualify for a new type, would say, "Don't worry, you'll have no problem." I said it myself many times in an effort to minimize a pal's forebodings. Though well intentioned, it was poor advice. We all had problems and we all worked like slaves in training, wondering each time if we had tackled something beyond us. When it got rough I'd remember old so-and-so saying I'd have no problems and wonder if he was not sharper and I more stupid than either of us realized.

One pilot, a more-than-competent airman on his own, was afflicted with checkitis to the point of near paralysis on a rating ride. A cooperative inspector agreed to a scheme to put him at ease, the fed riding along to rate a flight engineer. While the engineer solved his problems and answered questions, the instructor led the would-be captain through more practices of the maneuvers and procedures required to earn a rating, all of which the nervous fellow managed beautifully. After landing, the fed signed off both men.

Worry is corrosive. More than once, when my instructor said, "We'll go for your rating tomorrow," I thought he hadn't been watching closely enough, but he always proved to be the best judge.

With no prior experience whatsoever in turbojets, even as copilot, I reported for a 707 rating ride with considerable trepidation. The big jet was a handful; I was worried. The FAA examiner was dour, strictly business, and growled like a bulldog. I did the walkaround, answering questions about brakes, drain ports, and more of the scores of exterior items of interest on that awesome aircraft, which then it was in my view. We were walking toward the stairs leading to the cabin door when he said, "Your rating is in my pocket. Get up there and fly the way you flew it for your instructor yesterday." I did, and 90 minutes later he said, "That's all I need. Go sit in the back and let me fly us back home."

"Well?" I said.

"I'll sign it when we land," he said. The typical fed was not bent on busting a man, no matter how some regarded him. The process was like climbing a ladder, rung by rung, giving full attention to each phase in school and in the aircraft. If you thought ahead to air work while firing up, you could miss a low oil pressure or over-temping engine. Success lay in working deliberately, not hurrying through any procedure. A rating began with a clearance to a near-airport fix, to hold there. You read it back, explaining exactly how you intended to execute. Having outlined your plan, you might well be told to forget all of it before you were well established in the race track holding pattern. I found that explaining in detail how I proposed to solve a riddle sometimes worked to have its completion waived. Then came the "air work," stalls, steep turns, a wheelwell fire, emergency descent as though a cabin window had blown out, engine fire, and so on. Finally came a number of instrument approaches, all flown with one or two engines reduced to zero thrust as though they had failed.

You were expected to use your crew in all situations. Flying a large aircraft is not one-pilot work. The instructor played the role of competent copilot who would do whatever you asked him, but only if you asked. He was not permitted to prompt or remind. So you kept him busy tuning radios, reading checklists, and holding the control wheel while you reviewed approach charts. It was imperative to be completely in charge, to project the image of a skipper employing his crew to full advantage, being on top of every normal and abnormal situation. Hesitation and indecision would earn a thumbs down and dictate more simulator time.

Good thinking was expected at all times. If you broke out of the overcast at 100 feet, off to one side of the runway and 10 knots fast, you had 2 seconds to decide what to do with it. Maybe it could be ruddered back into the slot, crammed on and stopped on the runway, but rough maneuvering at low level raised questions about judgment. When in doubt, it was best to pour on the coal, pull up, and make another approach. A fed would usually allow a second try but never condone poor judgment, nor could he. He was not only examining technical skill, but gauging prudence in tight situations. His question — was the applicant adept enough, and mature enough, to safely operate this airplane with trusting ticket holders aboard?

A line check completed transition to the new type. This was a routine schedule flown with a check pilot in the right seat to start you off properly, a combination training and monitoring exercise, and then you were on your own — for six months as a captain. Copilot and flight engineer qualifications were good for a year. After which it was back to school for recurrent training, written exams, and a simulator check to see if you remained as proficient as the day you were rated. You passed or were held over for retraining until you did.

That was not the end of it. At least once a year a company check airman rode along in the cockpit for two trip legs to report on crew performance. A fed might appear anytime, unannounced, to warm the jumpseat and watch. On the paperwork would be a yellow slip reading, "FAA INSPECTOR JOHNSON AUTHORIZED COCKPIT DFW — JFK FLIGHT 4 THIS DATE." These federal observations usually went off smoothly, even with the occasional agent who considered some criticism to be a requirement of his duties.

"We're more closely watched than paroled convicts," grumbled a friend. Yet, for all of our bitching we approved of the system. I particularly endorsed it when I watched an airliner fly off into the night with my wife and kids on board.

15.

All in the Family

Some things you never forget. Looking back, they spring to mind as sharp as an event last month instead of something that happened 5 or 25 years ago. Such milestones are remembered with a grin, or grimace, all of your life.

This story began with an event more than three decades ago, at exactly 9:43 a.m. on a Friday, though I learned of it minutes later. "You have a daughter," said the nurse, peeling back the small blanket to reveal a doll-like sleeping face and incredibly tiny clenched fists. Five pounds, eleven ounces; that's not much girl, but I thought she was perfect. I still do.

We were soon put on notice that our household had been invaded and captured by a uniquely different personality. In many ways Kathy was as unlike her five-year-old brother as day and night. Terry was easygoing, adaptable, and possessed the wonderful gift of seeing the humor in every situation. His wee sister took one look about her and decided our relaxed style wouldn't do at all. She showed no inclination to fit in or to conform to any rules but her own. These were: meals right on the dot, fresh attire upon demand, and lavish attention — or complete privacy — at her whim. Within the week we were doing it her way and looking at each other, wondering what had happened. The fact that Dad had just tumbled into bed after a long night of flying his DC-4 counted for naught at breakfast time — her breakfast time, that is. With a howl of rage she would announce to the neighborhood that her 5:00 a.m. bottle was already five minutes late. "If she can't find other work, I'll get her on as a crew scheduler," I'd say, stumbling off to heat the water.

Son and daughter changed little while growing up. Basically, they remained the same Terry and Kathy. They were fortunate in having a mother who instinctively knew how to keep freedom and restraint in perfect balance. She kept our show on the road even when her mate was out on the line 10 days at a stretch. It was not an easy life for Margaret, no matter how our non-airline friends saw it. The airline dad is not always there to fix a leaking water heater, change a flat, or bus the kids across town for a game. He cannot rush home when Junior breaks his arm or Sis suddenly develops high fever. When there is a crisis, he's just leaving Tokyo for Manila and knows nothing about it.

To cut the mustard in the airline game, a wife has to be self-reliant, decisive, resilient, shrewd. Margaret was all of that and more. I remember arriving home dead tired from a week out flying the Pacific. As I put my suitcase down, the phone rang. Crew schedule — sorry to call you so soon but . . . "Is this what I signed up for?" she quipped, repacking my bag. Then, at the door, "Have a good trip, dear," and I was gone for another week.

There were many times like that yet she always made the best of them for the kids and me. "Daddy won't be here for your recital," I'd hear her say, or, "You'll have to give us a rain check again," punctuating her excuse with language used by drill sergeants. I once got a call from Texas while in a layover hotel in New York. After passing along family news — Kathy's birthday party was fun and Terry was in trouble with algebra again — she casually added, "I won't be home when you get in. I'm in the hospital. Well, I broke my leg — now don't get excited, it isn't serious."

The years rolled by. With Terry it was bikes, baseball, Cub Scouts with Margaret as den mother, Scouts, and vast collections of bugs, stamps, bottle caps, and plain junk. At one stage a plastic air force hung from his bedroom ceiling. This gave way to flying models, one or two of which did. Margaret thought she detected a talent that had escaped me, so we bought a trombone. When three lessons had not transformed our first-born into Tommy Dorsey, he lost interest and refused to practice. Have you ever tried to sell a month-old trombone? Or a set of drums?

Then it was ham radio with a transmitter built from a Heathkit and a receiver salvaged from an Army C-47. That was fun. We both qualified for licenses and the wall filled with cards from amateur operators all over the country. But it was with the camera that he hit pay dirt. He showed a genuine flair for photography. The radio station was dismantled, sold for a dime on the dollar, and the house thereafter reeked of the vinegary smell of developing chemicals. By bidding night freight runs I managed to keep him in Ektachrome without falling behind on mortgage payments.

His approach to school work could best be summed up as slap dash. The ability was there but not the incentive and no threats, promises, or explanations ("Sure I can fly to New York without the Pythagorean theorem, but you've still got to learn it.'') altered his conviction that school was a sinful waste of time. Oh, he passed, but by razor-thin margins at times. And my heart was not in the lectures. My old report cards, which he discovered, were evidence of my own distrust of Pythagoras who was in my view one Greek who should have been put away when he began drawing pictures in the sand. I silently agreed with Terry — who needed him?

He breezed through driver ed so we bought him a 1955 Chrysler. It was built like a Sherman tank, got the same mileage, and offered maximum protection. He never scratched it but his operating techniques were so often questioned by the constabulary that liability premiums became double the price of the car. ''You wanted a boy,'' said Margaret. Sometimes she was not much help.

''Let's give him flying lessons,'' I said. I had a theory about boys and airplanes. There was at a nearby grass field a quiet, patient instructor of 30 years experience who had a reputation for turning boisterous boys into good pilots. He agreed to try; I wrote out a check. Two weeks later, on a tip from good old Bert, we sneaked out to the grass strip and hid behind the hangar. The little Cessna rose with its single occupant, circled, and landed. Three times. It's hard to believe all that was so long ago. From then on, a car to Terry was but a way to get to where the airplanes were.

During the next two years he logged 1,000 hours, paying for most of it himself by doing airport chores, begging copilot time, ferrying new Cessnas and Pipers for expenses. He also managed to accumulate half the credits toward a college degree by signing up for courses he liked, a schedule which was the despair of his advisors. He began looking for an airline job, even buttonholing the president of our line though it had not hired a pilot's son in its 39-year history. His argument: my father says this is the best airline in the world; I want to fly for the best. It took more brass than I had at his age. Much to everyone's surprise, he got the job and entered DC-8 flight engineer training.

''You know you're on the spot,'' I said. ''You are an experiment. A lot of people will be watching you closely.'' ''Don't sweat it,'' was his casual retort, and I needn't have worried. He hit the books with a fervor which would have made him valedictorian at Thomas Jefferson High and graduated at the top of his class. Margaret and I could hardly believe it. ''But this stuff is important,'' he explained.

Meanwhile, Kathy developed at her own pace, in her own way. She thought dolls and play stoves were silly. Football was her bag. She was furious when forbidden to play with neighborhood roughnecks, particularly after tackling a bruiser twice her size with such enthusiasm his collarbone was fractured. She played all sports with strength and coordination far beyond her years. During the tomboy stage we bought her a sidewalk car with Briggs and Stratton engine. She would drive it around the corner, we were to learn, remove the governor and triple its speed. Then came the roller skating craze and, of course, she had to have a pair of those expensive high top shoes with wheels built on. She mastered all of the trick jumps and dance routines and was soon giving exhibitions for the frustrated parents of less agile offspring. It was her nature to give a new interest everything she had, master it, then drop it for something else.

She loved all parlor games but did not see them as pastimes. As in all things, she played to win. As merciless as a loan shark, she would quickly bankrupt the table with her hotels on Park Place and Boardwalk. School assignments were regarded as riddles to be solved with a minimum of fuss. She was hardly a scholar, caring little for most of the subject matter, but developed the uncanny knack of knowing exactly what would be asked on the exams. Her grades were consistently high and her name on the honors list, achievements she dismissed as of little consequence. There was something of the loner about her, the quiet restlessness of one whose eye is on distant, illusive, and private goals.

What a swirl of memories a daughter leaves behind — dance and piano lessons, corrective shoes and teeth braces, slumber parties, the first date, the first formal, wobbly high-heel shoes, hour-long phone calls, new clothes — unruly hellion one day, gentle heroine the next. She asked for, and of course got, a large trampoline. Not surprisingly, she was a natural at tumbling. I can see her yet, racing across the yard until dark doing flips, somersaults, cartwheels, staying with it until she had them down exactly right. It got her on as cheerleader in junior high, high school, college, and the Dallas Cowboys. "That was back when they wore clothes," Margaret emphasizes when that episode comes up now. It got Dad and Mom free seats on the 50-yard line at home games. It was something to hear that 75,000-voice roar go up when that youngster tumbled across the field like a circus pro, leading Dandy Don and his team to the bench. And something else to watch a nationwide telecast of a Cowboy game in a layover hotel as the camera focused on that familiar little five-foot blonde.

Then came the university, sorority life, and severance of the last strings, the last parent-child strings, that is, for we are closer to son and daughter now than ever before. You're happy to see them go out into life yet it breaks your heart when they do. Margaret must get the credit, having known when to hang on and when to let go. She was their best friend. She still is. Margaret welded four single-minded individualists into a solid family unit. "You should have written those books instead of Doctor Spock," I said. Her reply said it all, "Who's Doctor Spock?"

Every airline parent is asked, "Would you let your daughter be a stewardess?" and it became one we had to answer for ourselves. Late in her final college semester and with a good teaching job promised, Kathy said, "I want to try the hostess bit. What do you think?" It was never a question of our letting her do anything. She was her own woman and we would have had it no other way. But she wanted to kick it around with us which was flattering though not unexpected, so I told her what I thought.

An airline stewardess (we called her a "hostess" on our line, it being Tom Braniff's theme that passengers were guests on our flights) is the victim of much misinformation. She is called a waitress, a hatcheck girl, a barmaid, and, technically, she is all those. By the same token a nurse makes beds, runs errands, and empties bedpans and a first grade teacher is a baby-sitter who wipes noses. I see more to those vocations than the menial chores they entail. When we talk about the astounding strides made in our industry we dwell on technical advance, improved pilot skills, sprawling airports, forgetting that none of it would have come about had not the public bought the service. It may have been a magazine ad that enticed the businessman of 1940 to buy his first airline ticket, but it was service that brought him back. Fear of flying was the problem then and it was not only overcome by safety statistics but by the pretty girl walking the aisle of a DC-3, poised, efficient, friendly, able to set the nervous at ease. *She* was the airline to those first apprehensive travelers. She could make it or break it for the rest of us. What price can be put on her contribution and the goodwill reaped by tens of thousands of cheerful, helpful girls since?

And, while she has been good for the industry, it has been good for her. Many the timid, unsure girl turned into a confident, sophisticated young woman able to handle herself and her passengers in any situation. Overnight she becomes a tough, bright, gutsy gal you can depend on when the chips are down. I watched it happen a hundred times. It is hard work with little glamor to it, and brutally fatiguing at times. Yet it usually brings out the best as hard work does. The coffee-tea-or-me fiction does not demean her but those who believe it.

An airline can spend enough millions to buy itself an image. As the sole company representative present when the service is delivered, a hostess can confirm or destroy that image in the first minutes of flight.

For a certain combination of girl, the job can be the right combination of challenges.

We attended the ceremony and sat through the speeches. Terry and I pinned on her wings. Margaret beamed. We were the first flying father-son-daughter team in Braniff Airways' history.

I remember thinking, "Welcome aboard, Kathy."

16.

Primer for Passengers

We have come far since the DC-3 days, an era quaint in retrospect. Air travel was very much a novelty then, an exciting adventure to be relished and talked about long after. "I flew back from Detroit yesterday," had electrifying effect. You wore your Sunday best, were addressed by name, received the red carpet attention of a guest at Claridge's, and were in the company of the elite — and off you went at three miles a minute, smug in the illusion of superiority. You were somebody.

Today 20 times the riders fly together at three times the speed, one measure of the progress made. The technical progress, at least. In another sense we have retreated. In 1940 you bought your ticket and climbed aboard — period. There was no metal detector, no x-raying of suitcases, no armed guard staring to see if you fit the assassin's profile. It was assumed you were — well, ladies and gentlemen — you and everyone else in the terminal. There were no policemen to be seen.

The "airport security" now accepted as the normal prelude to air travel is depressing evidence of social regression. It is an affront to the 99.9 percent of us who observe the rules, yet its indignities are imperative. It is an expensive solution to a problem never dreamed of in DC-3 days. Some of your ticket money goes to pay the grandmothers who paw through your carry-on bags. The target is of course the maniac who commandeers an airliner; if he'll settle for a free ride to Havana, his fellow passengers are fortunate. This lethal moron has received too much press attention.

Airline crews wish nothing more than to make air travel a pleasant experience for the folks in back. That is what the job is all about. It's good business. One of my captains put it like this: "I don't work for the management, but for the passengers; they put the groceries on my table. The company collects their money and gives part to me." Delivering the quality of service every passenger has a right to expect is not always easy.

Rarely heard about beyond airline circles is the problem passenger whose irresponsible behavior makes life a misery for cabin attendants and everyone seated near him — or her — for the modern male has no monopoly on deplorable manners.

He's the inebriate who slips aboard unnoticed, or was slipped on by ground agents anxious to be rid of him, though this is strictly against company rules and federal law. The effects of his alcohol intake will likely increase as cabin pressure decreases so one of two things happen: he either collapses into noisy sleep or grows more boisterous. He may become belligerent. At best, he's an embarrassment, at worst a hazard.

Or he's Mr. Important, chock full of himself and his mission. Listening to his never-ending demands for special attention, you'd think he was riding his personal corporate jet. There's no pleasing this egotist. His sort erase whatever doubts you had about the "me generation." We carried our share of his kind. This pompous character invariably wanted the name of all the hostesses along with the captain's, mine because I didn't radio ahead to confirm his hotel reservation or advise his office we were running late. And he'd send up word that we'd all be in hot water when his complaint reached our bosses. One informed me that our president was a close personal friend. I let him find out for himself that the official he familiarly called, "Chuck," had been retired for two years. He is what the British so aptly describe as a crashing bore.

Mom and Dad join the list courtesy of Sis and Junior, unless their youngsters remain in their seats and keep reasonably quiet. We Americans are the worst offenders in this respect. Overseas families, particularly Orientals, put more stock in discipline and deportment. Few passengers buy tickets with the hope of being drafted as babysitters; in fact, every cabin is certain to contain one or two devout child-haters, and nothing ruins the day for harried cabin attendants like an unclaimed tot roaming the aisles.

And there's Lothario, straight out of Playboy's fashion section, magnetic, suave, sexy, utterly with it. Having purchased his self-image and swallowed whole the coffee-tea-or-me myth, he leers knowingly at the hostesses, not wondering if, but deciding which. After two martinis this clown can become disgustingly aggressive. It is indeed a mercy he cannot overhear the ruthless appraisal he inspires in the rear galley.

The airlines are to blame for some of it. Because their identical airplanes fly to the same places at the same speeds, fares and frills are the only differences to advertise. So they bill their trips as flying lounges, restaurants, and theaters with clear sexual overtones. Customers expect an open bar, silver service dinner, and an R-rated movie. The flying public has been led to expect too much. Blame the rest on the novel notion that rights need not entail responsibilities. I'd exchange the preflight electronic shakedown for the way business was done in DC-3 times. The good old days were not all that bad.

There are more incidents involving unruly riders than are generally known, some affecting flight safety. The carriers manage to avoid publicity in most cases and do not give their cabin crews the backing they deserve. An abused stewardess will not get far with a complaint unless she presses charges on her own, a move her management discourages. Indeed, press attention to an in-flight incident could land her in hot water. Nor are inter-crew relationships always what they should be, some pilots tending to shrug off problems behind the locked cockpit door.

If cabin crews called the shots, passengers would get a preboarding briefing, following the customary frisking for side arms and switchblades. It would go something like this:

"Your ticket entitles you to as safe, comfortable, and punctual a trip as our line can provide. All else is foam on the beer. Our prime concern at all times is your safety. That is the main reason we are here. We are well trained to prevent your injury in flight and to get you off the aircraft quickly in the event of a ground emergency or landing at sea. Keep your belt snugly fastened whenever the sign is on, and this means right to the terminal gate. Now and then, our pilots have to make quick stops during taxiing. If you're not secured, you could be injured.

"Beverages and a meal will be served enroute. Frankly, the service is too elaborate for a flight as short as this will be, being ordered by a vice president who never served a meal in his life, but we'll do our best. Everyone will be accommodated if there are no interruptions. Please wait for coffee refills until they are offered.

"In case you missed the notice at the check-in counter, please be reminded that interference with an airline crew member in the performance of his or her duties is a federal offense. You boys in mens' suits, go easy on the booze; you men in boys' suits, lay off the stews. Unruly passengers will be reported to the captain who can arrange an unpleasant reception upon arrival."

But you'll never hear anything like that! Crewmembers must solve cabin problems on their own. One of our hostesses found a way to handle the persistent troublemaker. Smiling sweetly, she'd whisper something in his ear. The effect is galvanic. The recipient would slump in his seat, wide-eyed, as though witnessing an unspeakable horror. He'd turn the color of putty. He'd seem comatose until brakes were set, then be first through the door.

She would never tell me what it was she said.

17.

Doing It Right

Every airline pilot should ride in back now and then. Leave the driving to someone else and be a part of what goes on behind the locked cockpit door. It's another world back there, and it's the world that matters for quality of service is the bottom line. Excepting fares, there are few other ways to outdo the competition.

Little things add up to make a flight nice, or just average, or downright unpleasant. The end result depends far more upon cockpit conduct than many pilots think.

On an evening we both remember, my wife and I flew to New York on a 727 nonstop, leaving at 6:00 p.m. The reservationist said the trip looked good for a pair of NRs (non-revenue riders), meaning there probably would be unsold seats. We employee family members only rode when space was available and there was no guarantee of travel until the doors closed. The landing gear doors, that is, for trips sometimes returned to the gate to exchange NRs for late-arriving ticket holders, such an outcome producing mixed emotions. You were upset by the disruption of personal plans, yet a full plane indicated a profitable operation and that's what paid our wages.

That evening we were lucky in being assigned the last two first class seats on Flight 12. We blocked out right on time, which is the best way to begin anything. No matter that tailwinds can recover lost minutes, it is still irritating to push back late. Passengers look at their watches, then out the window. An on-time departure suggests that the airline knows what it's doing, that it cares. It gets things off on the right foot.

The Three Holer could be an uncomfortable vehicle on the ground unless its driver (which is what the captain was at that stage) was careful. Easy through the turns, not too fast straight ahead, careful with the brakes — this fellow worked at it. On the way to the runway he made a brief announcement, introducing himself and crew, advising that we would fly at 37,000 feet through clear skies and arrive on time. And he thanked the fares for choosing our line, something most of us didn't do often enough. Saying thank you is not merely good

P.R., it's good manners. I felt guilty about my own P.A.s.

We rolled and rolled, slowly gathering speed, on and on. The temperature was a record-breaking 114 degrees that June evening at Dallas-Ft. Worth Airport and we were quite heavy. The departure was professional in every respect — flap retraction, power adjustments, pitch changes, turns. Smoothness is not built into an airplane. It reflects pilot concentration. Familiarity need not breed contempt. A pilot with genuine pride in his skill flies a load of freight with identical precision. Consideration for passenger comfort has little to do with it.

I had known that captain for a long time so his manner was no surprise. We rose above the summer haze and there was no way of telling when we leveled off for cruise. It's the little things that make big differences in air travel, the changes in flight path accomplished so subtly that the layman thinks his ride from here to there is flown in a straight line. Smooth flying pleases the knowledgeable who ride in back for it is the trademark of the professional; the pilot who stays ahead of his airplane stays ahead of everything else.

No sooner had Margaret remarked on our serene progress (she has few good words for air travel) than we ran into light chop — nothing much, but the seat belt light came on. In minutes we were out of it, during which I imagined a slight change in cabin pressure. Then this from the cockpit: we've descended to 33,000 feet to improve the ride, those lights on the distant left are Louisville, we're still on schedule. Nicely done, skipper.

Newspaper columnists like to jump airline pilots about their announcements. They should know that the spiels about seat belts, smoking, tray tables, and seat backs are required by law. Admittedly, some tour-guide chatter from up front is pretty awful. It is often too much, too little, or poorly delivered. Anything lasting more than a minute is usually excessive.

But, as pilots riding the plush always notice, passengers do pay attention to clear, concise, informative announcements. They are interested in where, when, why. The trick is in knowing when to say what, and how to phrase it. Many pilots never master it; some don't even try. Flight 12's had it down to an art. During three hours aloft he spoke three times, yet kept us completely up to the minute.

"You could take lessons from him," said Margaret — a cutting comment best ignored. But I resolved to do better on my own trips.

At its best, air travel is monotonous. At worst it is terrifying, particularly for the apprehensive who can find its sights, sounds, and motions fearfully unlike those of any surface travel. While airline crews properly focus their main attention to safe operation, the exercise is pointless with empty seats. The bottom line is too often ignored or forgotten. The quality of service which keeps seats filled is more dependent upon cockpit work than the length of the wine list. Airline crews should remind themselves of their own first airplane rides.

We slid across Baltimore and Philly in rapid descent and rolled onto final approach for a south landing at Newark. On the left, Manhattan Island floated between its sparkling rivers. The captain pointed out the George Washington Bridge, the floodlit Empire State Building and World Trade Center, then dropped us ever so lightly onto Runway 22L. He carefully taxied in and set brakes. We collected our things and walked through the tube into the terminal.

"What a wonderful trip," said Margaret, and I agreed. It had been as near-perfect as air travel can be. Neither of us was in the least influenced by the fact that the captain of Flight 12 that evening was our son.

18.

Christmas Out on the Line

Winter was near when the "frost-is-on-the-pumpkin" letter appeared in the reading file. That was about the first of October and we knew the old man had seen the TV forecast of the first cold front, what the locals called a "blue norther." In Dallas, that meant the temperature would plunge to 40, which *was* cold after a Texas summer. The gist of his two pages was that rough days lay ahead and we were to shape up, review cold weather procedures, and brace ourselves for ice, fog, and slick runways. The manual pages to review were cited.

It always began, "Gentlemen, the frost is on the pumpkin . . . ," and was always an embarrassment. You would have thought we couldn't read the calendar or had forgotten last winter and the one before and the one before that. It never dawned on him that crews flying to Minneapolis or New York had watched ice building on the windshield wipers for several weeks.

When the old man was kicked upstairs, a new man became the old man and his autumn letter began, "The frost is on the pumpkin," and again we were embarrassed. We went through a number of bosses, none of whom came up with a new lead. Airlines kept going in spite of chief pilots, and still do. But I poke fun unfairly. Such an admonition was required by policy. Its potential value was as an exhibit at the hearing if someone slid off the runway, or worse. "Yes, we caution our pilots. . . ."

That inane missive heralded what was in many ways the best time of the year. After Labor Day, the kids were back in school, Dad was out hustling orders, and our seats were filled. Which was the name of our game; full planes paid our bills. In October we came to grips with winter once again all over the system. "This is what it's all about," said one of my captains as we passed the outer marker at old Chicago Midway, following two trips which had missed approaches and headed for their alternates. We got it on the runway, hairily. Certain winter flying was grossly underpaid.

After Halloween came the heavy Thanksgiving loads. Take them home for turkey and, three days later, take them back to work. We ran extra sections and just about everybody got there in time, if not always on time. Then there was a breather while we got ready for the big one — Christmas.

Weekends and holidays had no special meaning out on the line, except that we were much busier before and after holidays, particularly Christmas. I always liked to fly to New York in December. Say what you will about the Big Apple — and you can't say much I'll dispute — it was a grand place at Christmas time, and I know it is still. The Yule hoopla back in Texas was too purposeful, too calculated to promote sales rather than to celebrate. I stood many times in the lobby of Grand Central Terminal and heard a magnificent choir singing carols, accompanied by an organ obviously set up at considerable expense. There were other entertainments offered at no charge and with no mention of sponsorship. Someone paid the bills but there was no apparent commercial aim. It was heart-warming to see such things done for their own sake.

And for the perfect supper on a wintry evening, nothing matched the incomparable clam stew served up in the Oyster Bar downstairs in that great station. Ian Fleming never dispatched James Bond through New York without that pilgrimage. Over the years the price rose from 95 cents a serving to $3.50, and there's no telling what it goes for today. Whatever the fare, I'd like a bowl right now.

Airport lobbies and ticket counters were decorated. Downstairs, where we went to study the weather and get the paperwork, there also were signs of the season — miniature trees, cards pinned up for all to read, elaborate teletype pictures depicting Santa or the Nativity scene laboriously composed at some out-station (the smaller the staff, the more elaborate its contribution), greetings from up and down the line and, on the 20th, a message from the company president to all employees. And, in the pilots' reading file, a letter from the old man thanking us for another good year and wishing us all the best for the next. That was from the heart and pleased everyone. We had it coming.

Airline life around Christmas was hectic; tempers were stretched, the atmosphere tense. Yet, as the Day drew close, people relaxed, made allowances, gave the benefit of the doubt. You could feel the change. It was a good time to be out on the line. I remember handshakes and friendly exchanges between individuals who routinely ignored each other, even detested each other. I remember single pilots with Christmas off offering to swap trips with dads scheduled to fly. You don't forget gestures like that or those who made them.

I remember the radioed exchanges between crews and the good FAA people who worked in towers and airways control centers.

"So long, O'Hare Tower, you all have a Merry Christmas, you hear?"

"Thanks, Braniff, and you, too. Say, do you guys have tomorrow off?"

"Affirmative."

"I wish I had your seniority. I've got to work."

"Maybe it will be next year for you."

"I hope. Okay, call Chicago Center on 131.5 and tell 'em Jake sent you. So long."

And, if you did have to fly on Christmas Day — "Welcome to Chicago, Braniff, and a Merry to you, sir. Now, if you'll reduce your speed to 200 knots and turn right to 75 degrees, I'll try to squeeze you into Runway 4 Right. Can you make it down to 2,200 feet at the outer marker?"

"Thanks, and a Merry Christmas to you, sir. We can do that and we appreciate the short cut."

One frigid Christmas Eve we crossed Nashville just before midnight, riding smoothly in the clear. The speakers had been silent for minutes when was heard, "Attention all aircraft. This is good old Eastern Air Lines Flight 26 wishing everyone on the frequency a very Merry Christmas." There were a dozen responses from competitor trips and the radar man monitoring our blips on his screen in Memphis Center. Good old Eastern, yes indeed.

On board, the cabin crew did its own decorating and always reminded the cockpit to make the traditional Christmas Eve announcement. "Ladies and gentlemen — *and boys and girls* — we have just heard from ground radar that a strange aircraft is southbound from the North Pole. They say it's a sleigh drawn by eight tiny reindeer. We will land in Dallas two hours before it gets there."

It was during one Christmas week that we flew the final leg of the long haul from Vietnam to California. In back were 165 Marines, subdued as all returnees were. They sat and looked at nothing, lost in their own reflections. Riding our cockpit jump seat was a deadheading copilot, a huge fellow weighing 220 pounds. When we leveled off, he unpacked a complete Santa Claus outfit, white beard and all, and donned it without a word. Then he walked through the cabin, handing out peppermint sticks to every man. That stunt took some planning. Did it go over? I could hear the cheers from the cockpit. The senior cabin attendant, a tough girl who'd seen it all in 25 years came forward, shut the door, started to light a cigarette, then wept. It was one of those trips you never forget.

I remember flying southwest on Christmas Eve with the cockpit stacked with presents bought at Macy's and Gimbel's riding through the moonlit sky across the snow-covered landscape, going home for Christmas and thinking of what the kids would say when the boxes were opened. And knowing that I'd be up until very late fitting Bracket R7 into Slot W9. And wondering what Margaret would say when the American Express bills came in.

We blocked in, dropped our brain bags in Operations, and hurried to our cars.

"You fellows have a Merry Christmas, both of you."

"You, too, Captain, you and yours."

"See you in a couple of days."

"We'll be here."

Christmas was a good time to be out on the line.

19.

The Abort

Reward a pup for coming when you whistle and soon enough it will come, reward or not. That is a conditioned reflex. Developing the proper conditioned reflexes is what good flight training is all about.

Consider the classic exercise for multi-engine pilots: the complete loss of power from an engine during takeoff at maximum gross load. The trainee at our shop didn't lose an engine during departure most of the time, but every time, and he lost it just prior to liftoff. He had all engines turning perhaps 10 percent of the time spent in airplanes and/or flight simulator. Four-engine candidates spent part of the remaining hours with two out, usually on the same side. Holding in enough rudder to offset the yaw from asymmetrical thrust was equal to jogging in restoring tone to leg muscles.

Back in DC-3 days, a cloth over the throttles kept you from seeing which one the instructor pulled. You identified the failure with your feet; hard rudder was the good side.

There was more to it than keeping the unwieldy beast out of the trees. You had to adhere exactly to the assigned departure routing, oversee engine shutdown, dump fuel to reduce weight, and request clearance to return to the airport, to which the invariable response was, "The field is now reporting zero-zero. You are recleared to Hard Nose VOR, to hold east. Maintain 4,000." You were busy. It didn't matter that the training aircraft was empty; reduced takeoff power produced the same sluggishness as a full load.

The earliest multi-engine crews got no such indoctrination, their machines being unable to overcome the drag of a windmilling propeller. British Handley Page 0/400 bomber pilots in 1918 were told to put it down straight ahead if one packed up, regardless of what lay ahead. It was their best chance. The monstrous German DO-X flying boat 10 years later settled earthward with but one of its 12 engines dead. Such early American airliners as the Lockheed 10A *Electra* and Douglas DC-2 could struggle along for a limited time with one out, providing the terrain-weather-load equation was favorable, then came the DC-3 with full-feathering propellers. It could keep going indefinitely on half power if handled with a watchmaker's touch. Its crews were required to master the tricky procedure, the basics of which have not changed since.

So you went through it over and over until you had it down cold. And you went back to school twice a year to show you could still cut the mustard. If you advanced to a new type — or retreated to an old one — you went through it all over again. It was heartening to learn that an airplane could climb above obstructions if an engine failed as the nose was lifted, for it had happened and could again, perhaps to you. Of course, you didn't really think it would happen to you. At least, I never did. What, lose an engine at the critical instant of rotation when weighing the maximum? The odds against it must have been in the millions.

But it could happen to me, and did. Whatever subconscious comfort the mathematical odds provided was forever wiped from my thinking one hair-raising afternoon at Honolulu. I had just completed 747 training — groundschool, a week in the simulator, an hour and a half in the ship shooting approaches and landings. Twenty-five hours on the line with a check airman would wrap it up if I did well, otherwise he would remain in the copilot's seat until I did.

Sent along to look me over was Jay, a good friend and fellow pilot for 30 years, but don't assume anything from that. Friendship counted for naught on checkrides. A company man was always harder to please than a fed, which was as it should have been. The average airline demanded more of its crews than federal minimum standards. The best lines still do. Jay was shown on the paperwork as pilot-in-command though I warmed the captain's chair.

A questionable engine indication caused us to discontinue the takeoff from Dallas-Ft. Worth Airport after reaching 80 knots. We taxied back, did some double-checking, and broke ground 30 minutes behind schedule. The trip westward was a piece of cake. We blocked in at Honolulu eight hours later. So far, so good.

A reroute message was waiting: due to a typhoon nearing Guam, we were to bypass it the next day and fly nonstop to Hong Kong. No problem.

Aircraft N610, one of the first 747s built, was assigned to our Flight 505. Lufthansa had worked it hard for nine years before we got in on lease. We pushed back with 180 passengers, 17 crewmembers, underfloor bins filled with freight and full tanks. Takeoff weight: 709,960 pounds, 40 pounds below the legal limit for the runway and temperature of that moment. We taxied slowly out to Runway 26L, 12,001 feet long with a 560-foot blacktop overrun on the far end beyond which was a steel fence and the vast Pacific.

As we swung into position on the runway, I mentally reviewed the takeoff and engine-out procedures, never dreaming I was about to put them to the test. Our book read like this: the pilot flying — in this case, me — was to keep his left hand on the control wheel and his right on the thrust levers until the second pilot called out the V-1 speed, at which time he placed both hands on the wheel and began with slight back pressure to lighten the nose. The V-1 for our specific situation was 155 knots — 178 miles an hour. At a slightly higher velocity called V-rotate, the nose was to be raised to eight degrees and, if this was smoothly accomplished, the airplane would lift and rise to 50 feet by the time it reached V-2, the speed upon which climb was based.

During certification, Boeing had demonstrated to the FAA's satisfaction (as every plane builder must with a new airliner design) that its jumbo, fully grossed, could be brought to a full stop "within the confines of the runway" if an engine died prior to reaching V-1 speed. The procedure: chop all power, pull the speed brake handle to raise the wing spoilers, reverse symmetrical engines only, brake hard. If the failure occurred after V-1, you flew, the drill being to hold the nosewheels firmly on the ground with forward pressure on the wheel to insure directional control until V-rotate, counteracting yaw with the rudder.

When thinking through the performance data supplied with any airliner, I always found three things slightly disturbing: 1 — the qualifying tests had been conducted in a spanking new machine; 2 — the professional test pilot who accomplished the "demonstration" knew beforehand exactly what was going to happen; 3 — he was probably a better man than I. I wondered if the feds added a fudge factor to compensate for utter surprise, not to mention airplane (and pilot) age. And they had. The test pilot had not been allowed to use reverse thrust and had to bring his ship to a halt using brakes only. I still leaned heavily on the odds against it happening.

And there we sat on that sparkling windless day, waiting to go. Cleared for takeoff, said the tower. I eased the thrust levers forward an inch, waited for the four rpms to stabilize, released brakes, and worked the big handles right on top until maximum thrust showed on the gauges, feeling the flight engineer fine-tuning them beneath my grip. The enormous Pratts took up their awesome thundering growl and we began to move, all 355 tons of us, very slowly at first.

My airspeed needle began to move, inching through 60 knots, 80, 90, 100. Due to the 747's cockpit height we seemed to be creeping along. This bothers every new jumbo pilot, but now the needle pointed to 120 no matter how it looked outside, then 140. My final glance at my gauge saw the needle pointing to 150 knots. I was moving my right hand from thrust levers to wheel when it happened — a sharp, hollow boom as though someone just outside had fired a shotgun. The nose wrenched savagely to the right. Jay snatched the thrust levers back and yanked the speed brake. I ruddered us straight, pulled all four handles into maximum reverse, completely forgetting the symmetrical-only procedure, and stood on the brakes. Now, with Numbers 1, 2, and 3 in reverse, the nose slewed to the left.

The view ahead was not encouraging. We'd consumed a lot of concrete building speed. Looking at the remaining runway, the blacktop overrun and the sea beyond, and feeling the monster's reluctance to decelerate through my hands and feet, I thought, we're not going to make it. We're much too fast, far too heavy, there's no way. For a fleeting instant it occurred to me that the passengers in the cabin below would get wet before we did. But . . . we *were* slowing. There was 110 knots on the gauge, then 90. It began to seem as if we might keep the airplane dry. We rolled ponderously past the end of the runway and onto the black overrun, then slowed to a walk, and finally ground to a stop, the nosewheels 18 feet from the far end of the blacktop. And thus ended the longest two minutes of my life.

A hostess dashed into the cockpit shouting something including, "A fire!" One of us, I don't recall who, pulled the Number 4 fire handle and fired both bottles of extinguisher. The engineer threw open the upper deck side door and reported no flame or smoke from the lost engine. We shut down Numbers 1, 2, and 3 and just sat there. We told the chief hostess that, no, we would not evacuate down the slides but to leave them armed just in case. The ship was listing to the right, someone noted. Not much else was said. It was enough just to sit and look at all the sparkling blue water beyond the nose.

The Number 4 turbine had distintegrated, engulfing that engine and right wing tip in a single flash of flames, which was what the chief hostess had seen. Our not-by-the-book asymmetrical reversing (without which we could never have remained on dry land) had required harder braking on the right, causing fuse plugs in the eight main wheels on that side to blow. All the left main plugs blew minutes later. The treads of the two brand-new nosewheel tires had been completely scrubbed off, exposing the casings. Otherwise, we were intact. By a fluke we had not hit a single runway light standard and had even rolled between the red lamps marking the far end. Pure luck. The tire marks were visible a year later and may still be there for all I know.

It made the morning paper with a front-page picture and typically inaccurate story blaming the incident on tires blowing out. No comment from us. We saw the newspaper and TV cars racing toward us behind a police cruiser. The airport manager rescued us as they pulled up beside the silent abandoned 747, and drove us away. The reporters wouldn't have understood what conditioned reflex had to do with it.

20.

Reunion

Years ago I set about compiling my airline's fleet list, that is, tracking down every airplane that had flown under our logo. When was it bought or leased, how long did we operate it, and where did it go after we sold it? As any researcher could have told me, I had bitten off a mouthful.

The jets were easy, though one 727 was a problem in that the same manufacturer's serial number appeared against two tail numbers. I eventually determined that we had leased it, returned it to its owners, then bought it under a new registration. Registrations are often changed, builders' serials never. Going back through the *Electras,* DC-7s, -6s, Convairs, and -4s posed few problems. The DC-3s were something else, with ships being taken over by the Army during the war, acquired through merger, or bought as surplus military equipment.

We Americans do not keep particularly good records despite our mania for paperwork. Our corporate files were incomplete and inaccurate; FAA records proved little better, many having been lost in a fire. British enthusiasts are extremely good in this area, however. A tower controller in Scotland who had never been out of the United Kingdom filled in most of the blanks, though how he got his information I have no idea. During our exchange, he asked for the tail numbers of C-47s I had flown during the war. He wrote back that, having matched them against his cards, I would be interested to learn . . . but that's getting ahead of the story.

Thirty years earlier, events put me in Accra, Gold Coast, a sweltering place 400 miles north of the Equator. It was there that transport and ferry traffic crossing the South Atlantic made African landfall. Under contract, Pan Am was building a terminal which shortly would surpass anything back home in size and capacity. The aircraft at that early stage were mostly DC-3s commandeered from the airlines; the C-47 version with stressed floor and cargo doors was just coming off the assembly lines. I would be a copilot in the ambitious plan to span

Africa with scheduled transport service. Accra was Headquarters, African Middle East Wing, Air Transport Command. An hour of Link and two in the airplane comprised training. Back then a copilot was expected to trail the cowl flaps before takeoff, operate gear and flaps enroute, and absorb wisdom by osmosis. The first ride was in a spanking new C-47, tail number: 17223.

"You will be interested to learn," wrote the Scot, "that the *Dak* (*Dakota* was the RAF name for the C-47) in which you trained is now displayed in Arizona." Well! I had to see it again.

There is no such thing as a bad air museum although even the best are somehow depressing to a pilot. You remember an aircraft making its characteristic sounds, moving, flying, as something alive. Yet, far better a static display than to let a worthy type disappear, the sad fate of too many great airplanes already. And, it is true that a number of rare specimens kept in flying condition have been destroyed in accidents. Pima Air Museum at Tucson, with 110 aircraft, is the third largest collection in this country. Though primarily made up of machines obtained from the Military Aircraft and Disposition Center at nearby Davis-Monthan AFB (present aircraft population: about 4,000), there is a surprising variety of types to be seen close up.

Among the 26 fighters is an F6F-3 *Hellcat* which rested for a quarter of a century on the bottom of the Pacific before being recovered by the Navy for a study of corrosion effects. It is displayed as-is, and it is in remarkably good condition, considering. A P-38 *Lightning* and P-63 *Kingcobra* are among the few World War II machines present, the Museum not being established until 1966 after many thousands of notable airplanes had been melted down. Most are jets, among them three which for some reason always seemed as if they'd be great fun to fly — a Lockheed F-94 *Starfire*, a Grumman F-11 *Tiger* (this example once flown by the Blue Angels and still in air show colors), and a Lockheed F-104 *Starfighter* (the notion of strapping on that tiny wing shoved along by that much engine makes me breathe faster).

The 26 bombers are what you'd expect. Among the most interesting to me are the Douglas B-18 and B-23, neither boasting exciting records, but well-remembered types from my behind-the-fence boyhood. The B-24 *Liberator* is of special interest in that it was donated by the Indian Air Force and ferried to Arizona in short and, by all accounts, memorable stages. The B-52 *Stratofortress* on display is the one used to launch the X-15, those historic drops being noted on its flanks. I like museums that provide specific information on each exhibit. The USAF Museum at Dayton follows the same policy.

Among the attack aircraft is a North American RA-5C *Vigilante* which you will pause to admire if you ever watch one fly. How that supersonic monster was brought to rest on a carrier's deck is beyond me. (And did North American ever build a bad-looking airplane?) One of the three North American F-107As built is there (it lost in competition with the Republic F-105) and a full-scale mockup of the X-15. And there are plenty of helicopters, pilotless missiles, and flying boats for anyone interested.

Having spent my life in the relatively staid sphere of transports, I was particularly interested in the Lockheed 10 *Electra*, a small twin operated by my airline in the 1930s, and the Boeing *Stratoliner*, a remarkable airliner in its day and a hint of things to come from Seattle. I rode aboard one in 1950. Through a thunderstorm. A Lockheed *Constellation* still in the colors of its donor, TWA, stands close to the Army version, a C-121A, this one being the personal aircraft of General Eisenhower as commander of SHAPE in postwar years.

Near the entrance is a USAF model of the Douglas DC-6, this one a VC-118A employed for a time as Air Force One for Presidents Kennedy and Johnson. This is an exhibit you can get aboard. A visit is worth the effort if only for the guided tour through this historic airplane. It has been restored to its original VIP configuration, right down to the red phone.

I eventually came upon my C-47. It's been repainted in D-Day stripes and now has a Kilroy cartoon under the left cockpit window, but there on the vertical fin stenciled in yellow is "17223." I wondered where it's been since that humid morning in 1942. Pima Air Museum officials also would like to know. They got it from a row of C-47s stored at Davis-Monthan and know little of its history except that it was the second production C-47 off the Douglas line. I seem to recall my Scot friend saying the French Air Force flew it after the war, then, after whatever other duties it fulfilled, it served with the Wisconsin Air National Guard — but I'm not sure of that. Someone knows.

Across the road from the Museum are several aircraft salvagers dismantling ex-military equipment for parts, among them several C-47s. The carcasses go to the smelters and will ultimately end up as beer cans, I guessed. "No, most of it ends up as trim on automobiles," said the manager of one yard. I asked about a control wheel and was shown a pile of columns with wheels still attached. Mine is in the den, mounted alongside other sticks and wheels of airplanes once intimately known. I can see it from where I sit.

Once in a while I grip that slender wheel and remember all sorts of places and people and things that happened, some good, some bad.

21.

Back Yard Air Force

After moving into our rural home in Texas, we settled into the comfortable routine of rising at seven to drink coffee, read the paper, and watch the birds. It was a most pleasant way to start the day. The large back windows afforded a fine view of the yard and pond at the foot of the slope. Yes, I knew that big windows are not energy efficient when I drew the floor plan, but the view was well worth the expense. You pay for what you get. We were never the types for an underground house.

Lest you assume that approaching retirement had sent me around the bend, as the British say, let me add that my bird watching goes back a long way. Many the lazy boyhood hour was spent lying in the summer grass on a Georgia hillside, admiring the ineffably graceful flight of buzzards. Loathsome creatures though they might be when gorging on the entrails of a dead animal, there was a real beauty in their lazy wheeling in the thermals.

I turned in a third-grade essay on buzzards; it drew a low grade and much joking. I still admire the buzzard and still watch him. That fellow can fly. Georgia was a good bird place. Texas was even better; in fact, it is the best of all the states with 540 species recorded. California runs a poor second with 80 fewer.

I cannot fathom a pilot who has no interest in birds. After all, it was fascination with birds that led to the first secret dreams of human flight.

When we lived in suburban Dallas a surprising variety of feathered visitors were attracted to our back yard with a feeder, bird bath, and a dozen wood and tin-can houses. English sparrows took over most of them after huge free-for-alls for possession. The sparrow is a feisty little street fighter, a city fellow. He's at ease around people and tries to communicate, always with demands. One day six of them lined up on a limb just outside the window, looking in at us, chattering wildly. We saw the cause, a gray cat half hidden in the bushes. Monty took care of him. In most respects a genial old boxer gent, Monty (for the general) harbored a blind hatred of felines.

Another time, a sparrow and his bride noisily called for our attention. The wind had blown their house askew and they were complaining to the landlord. I righted it, taking care not to disturb the chicks inside. Pop and Mom moved back in and that was that. Some neighbors used air rifles to chase sparrows away. Not us.

A family of cardinals nested in the back hedge, but to most visitors we were Holiday Inn, a place to wash up, eat supper, and rest overnight. They lived elsewhere. Hummingbirds occasionally were seen around Margaret's flowers, but they ignored the red sugar water we offered. Each fall vees of geese and ducks crossed, way up there, but there were lots of mockingbirds, robins, thrushes, and woodpeckers to watch close up. And jays, those loud-mouthed flashy dressers whose bullying manners failed to rattle most other birds. A jay foolish enough to perch on a sparrow's housetop had bought himself a scrap.

Then we bought three acres 35 miles north of the new DFW Airport and built the country place. I put up the antique bird bottles bought at Williamsburg and ten boxes, wondering if there was a demand for housing on a lot with hundreds of trees. Sparrows seized the bottles but the wood houses were ignored, at least I thought they were until I inspected them the next spring. Each had a nest. One held a real surprise, a five-foot water moccasin which had wintered there. The same box was the winter home of an even longer king snake the following year, non-poisonous but capable of a painful bite. Thereafter, I cleaned out the old nests carefully, shotgun nearby.

At first, the feeder fell flat. They'd all sit in the trees looking, discussing its temptations, then fly away. Birds are so much like humans; the gullible soul is more often the city dude than the rural hick. Country birds are wary. One morning a gutsy sparrow lit on the feeder, eyed us with suspicion, and filched a scrap of bread. That did it; there was no keeping it stocked after that. Again, the similarity with people, the timid hanging back until a single brave soul gave it a try. You might know it would be a sparrow.

The hummingbird bottle of colored sugar water was enormously popular. The first guest had a green head and red collar; he lunched leisurely, paying us no heed, and was soon back with his girl friend, standing watch in a tree while she fed. A variety of the tiny birds entertained us all summer. One morning in early fall a pair stopped by for a snort, then hovered side by side just outside the window looking in at us for a full minute. "Adios, amigos," was clearly the message for we saw them no more that year. They were Guatemala-bound.

One morning the following March there was a flurry of activity at the window. There he was, green head and red collar, hovering before the exact small pane where the bottle had hung. I put out the sweetened water, dyed red. He took one sip then sat in a tree staring at us. He was still there at noon, waiting. I mixed a fresh batch with more sugar. He sampled that, gave me a sharp look and went back for a long lunch. I should have known better than to cut the drinks on a fellow aviator sharp enough to navigate 1,700 miles and find a certain pane of a certain window of a certain house in north Texas.

You would think the expenditure of energy required to hover would make them short-range fliers, dependent upon frequent refueling, yet they are known to cross the Gulf from Yucatan to Louisiana, 800 miles, obviously flown nonstop. Though weighing as little as a tenth of an ounce and delicate enough to be trapped in a cobweb, the tiny creature is absolutely fearless. He will attack an eagle 1,600 times his weight if one invades his nesting territory. Although I've never heard *Hummingbird* suggested as the name for an interceptor aircraft, no name could be more fitting.

Our pond attracted spindly-legged egrets, huge ungainly birds when wading near the bank, fishing, but what great beauties in flight. Those jumbos soared across the water inches high, apparently using the ground cushion just as pilots do when landing.

Mallards paused in their migration to spend a night or a week on the pond, but only after circling several times at high speed to check us out. Curiously, they always seemed to follow left traffic, as do pilots. Their flight was spectacular but their landings were something else. The typical final approach was right on the money, a huge splash, and then it was butt over tea kettle. Earlier arrivals took no notice, possibly because they could do no better.

There was a roadrunner that lived down in the woods. He'd come around once a week, approaching in characteristic mad dashes and sudden freezes, hop up on the sun deck, and peer in the windows, curious but extremely shy. When startled, he flew with remarkable speed and grace.

I erected a 12-bedroom martin house well away from the house and trees, just as the instructions said. Sparrows took over two rooms before I got the ladder put away. Twice I lowered the apartment and removed their furniture, hoping they would move to one of my nice single-family condos. But no, they preferred communal living. I gave up. You could never win an argument with a sparrow. Sure enough, right by the book, three martin scouts showed up, inspected each apartment, and left. They were back in a week with the rest of their clan, moved in, and lived in harmony with the sparrows all summer.

The flight of the purple martin was something to watch — diving and climbing as though exhilarated by the miracle of flight, though we knew it was strictly the business of catching insects on the fly. He's very good at it, too; we never saw a mosquito or, for that matter, many other insects. We took care of the birds and they took care of us. I never tired of watching a martin returning home. He'd dive at the base of the pole, pull up into a wings-folded vertical zoom and run out of speed just as he reached the porch before his door. The timing was always perfect. You can't do that in a plane.

A pair of mourning doves sat in a scrub oak right beside the house, admiring each other, smooching, cooing, discussing house plans. According to our bird book, the mourning dove is a lover first and a house builder second; in fact, it is noted for slipshod building construction. This proved to be entirely accurate. Between long expressions of undying love, the pair put together a flimsy arrangement of twigs atop a large limb with no cross-bracing whatsoever. Sure enough, the first breeze blew it off, and we never saw them again.

Then we moved to North Carolina, to an old farmhouse on 12 acres. I built two boxes and mounted them in trees outside the sun room windows. One was immediately taken over by a pair of lovely bluebirds, a breed we rarely saw in Texas. The other soon contained a nest, but we didn't see who built it. A feeder filled with sunflower seed has been popular with a variety of finches and other small birds, but the state bird, the cardinal, simply sits in a tree watching. Texas cardinals used to monopolize the feeder, daring anyone else to join them. Strange. I bought another Sears, Roebuck martin house and put it up well clear of trees. We haven't seen a purple martin yet, but it was not long vacant. It was immediately taken over by sparrows.

22.

The Rules

Every profession has it mottos, adages, axioms, and principles. Here are some from mine. Whether or not any can be termed a Great Truth depends on where you sit:

Flying is not dangerous; crashing is dangerous.

An airline pilot is a confused soul who talks about women when he's flying, and airplanes when he's with a woman.

Any comment about how well things are going is an absolute guarantee of trouble.

A captain is two flight engineers sewn together.

Clocks lie; an 18-hour layover passes much quicker than an 8-hour trip.

If it ain't broke, don't fix it; if it ain't fixed, don't fly it.

Crime wouldn't pay if the FAA took it over and would go bankrupt if an airline management did.

Winds aloft reports are of incomparable value — to historians.

Any pilot who relies on a terminal forecast can be sold the Brooklyn Bridge.

A grease-job landing is 50 percent luck; two in a row are entirely luck; three in a row and someone's lying.

The more traffic at an airport, the better it is handled.

The friendliest stewardesses are those on the trip back home.

The longer the schedule, the greater the odds of having an inoperative autopilot.

There are four ways to fly: the right way, the wrong way, the company way, and the captain's way. Only one counts.

Flying is a vocation for men who want to feel like boys, but not for those who still are.

Arguing with your captain is like arguing with a radar cop.

Jokes told by flight engineers are ignored; copilot jokes draw smiles; captain jokes send copilots and engineers into hysterics — none of which proves who tells the funniest jokes.

A captain with little confidence in his crew usually has little in himself.

An airplane may disappoint a good pilot, but it will rarely surprise him.

The sharpest captains are the easiest to work with.

A control tower tape that supports your version is certain to be "accidently erased."

You can always tell an airline pilot, but you can't tell him much.

Nothing is more optimistic than a dispatcher's estimated time of departure.

A four-time loser: the fellow who went to Texas A&M, joined the Marines, flew helicopters, and was hired by Braniff.

The owner's guide that comes with a $500 refrigerator makes more sense than the one that comes with a $50 million airliner.

The odds of having a check rider on the jump seat decline with the destination ceiling.

If it doesn't work, rename it. If that doesn't help, the new name isn't long enough.

"Please see me at once" memos from the Chief Pilot are distributed on Fridays after office hours.

Flying skill and administrative talent are not often found in the same person, thus the average chief pilot is either right where he should be, or he's a total misfit.

Everything in the company manual — policy, warnings, instructions, the works — can be summed up to read, "Captain it's your baby."

Most airline food tastes like chicken because most airline food is chicken.

If an earthquake opened a 10-foot fissure in the runway that caused a landing mishap, the National Transportation Safety Board would blame it on pilot error.

One hole in the overcast is worth 10 published approaches.

A thunderstorm is not usually as bad on the inside as it appears from outside; it's usually worse.

When discussing yesterday, the weatherman is a scientist; when talking about tomorrow, he's reading tea leaves.

A good check ride in a flight simulator is like a successful appendectomy on a cadaver.

Out on the line, all the girls are looking for husbands and all the husbands are looking for girls.

A groundschool instructor understands piloting the way an astronomer understands the stars.

The level at which turbulence will be encountered is readily determined using the formula: assigned cruising altitude minus 500 feet.

The only thing worse than a captain who never flew as copilot is a copilot who once flew as a captain.

A copilot is a knothead until he spots opposite direction traffic at 12 o'clock, after which he's a goof-off for not seeing it sooner.

Any pilot who does not privately consider himself the best in the business is in the wrong business.

Every groundschool class includes one ass who, at 5 minutes before 5, asks a question requiring a 20-minute explanation.

Were airline operational philosophy applied to the entertainment industry, Baldwin Piano Company would tell Joe "Fingers" Carr how to play. If he dared his own interpretation of "Tiger Rag," he would be fined. If he did it again, he'd be laid off for 30 days.

Only one thing is thinner than the ham in an airport sandwich. The cheese in an airport cheese sandwich.

The only soul more pitiful than a captain who cannot make up his mind is the copilot who has to fly with him.

Federal Aviation Regulations are worded either by the most stupid lawyers in Washington, or the most brilliant.

Standard checklist philosophy requires that pilots read to each other the actions they perform every flight, and recite from memory those they need once every three years.

The most sensitive mechanism in modern aviation is the shower control in a layover motel.

Everything is accomplished through teamwork until there's an accident, then one man gets all the blame.

If it's lousy here, it's probably good where you're going.

An attempt to stretch fuel is certain to coincide with an increase in headwind.

A crew scheduler is the sort who would wake his wife at midnight and order her to carry out the garbage. Then send her back for the cat.

The most fascinating topics in aviation are the people who design, build, fly, and maintain aircraft, yet nearly all air literature is about the machines they design, build, fly, and maintain.

Anyone can fly trips, but it takes a genius to bid trips.

Tell strangers you fly for another airline and they'll tell you how much better yours is.

Jet and reciprocating engines work on exactly the same principle: suck and squeeze, blow and go.

The louder a copilot complains about his skippers, the more likely he will become an insufferable captain himself. The engineer who gripes about his copilot and captain will be a pain in the neck in both pilot seats.

The pilot who plays banker when settling the tab for a crew dinner will come out six dollars short — and be expected to leave the tip.

An accident investigation hearing is conducted by non-flying experts who need six months to itemize all the mistakes made by a crew in the six minutes it had to do anything.

The most nerve-wracking of airline duties: the flight engineer's job on a proving run flown by two chief pilots.

It is easier to cope with a single major in-flight problem than a series of minor ones. Real trouble must be swallowed in small doses.

The hotter the layover date, the greater the chances of arriving four hours late.

A good captain and copilot go hand in hand — but not through the terminal lobby.

23.

Lasting Impressions

Four decades of planes, places, and people left their impressions, good and otherwise. Some are based upon brief encounters, others stem from lengthy associations. The aircraft types in my logs do not comprise an impressive list, and although I can connect interesting dots on a world map, vast regions remain unmarked. So much for disclaimers. These thoughts stick:

The little Fleet 16, known to us in the Royal Canadian Air Force as the *Finch*, was the most wonderful airplane ever. On a well-remembered summer morning in Ontario it bounded across a clover field — its five-cylinder Kinner engine barking importantly, lifted me into the airman's realm, and fulfilled a boyhood dream. A ride in the space shuttle could not provide the same thrill. The Fleet was a little doll.

Then came the *Harvard* (known to our Army and Navy as the AT-6 and SNJ), a howling, hairy-chested handful of flying machine, at least for a teenager. The perfect trainer, it taught me much about flying and about myself. Flying brings out your best and worst.

Harvard grads normally went on to fighters but I was assigned to slow transports, a crushing disappointment at the time. The unwanted assignment certainly saved my life; an appalling percentage of friends in fighters were soon dead. Eventually, I appreciated the value of our contribution and enjoyed the duty, which was just as well for I was stuck with it for the duration. After the war I flew types well known to every veteran who found an airline berth.

The DC-3 was loose-jointed, flew as if tied together with rubber bands, accomplished the impossible, and was without question the most important aircraft ever built.

The C-46 *Commando:* big, ungainly, and put together like a brick privy, it handled like a Peterbilt with front-end problems but was a workhorse equal to any task.

The DC-4: wonderful in every way, beautifully engineered and so easy to land that it made you look better than you were.

The DC-6 was pressurized and cruised at 25,000 feet. It was a good, honest ship with stiff legs that made consistently smooth arrivals impossible.

The Convair 340/440 series: noisy, dependable and versatile, it worked its crew hard. If you could fly it well, you could fly anything that came later.

The L-049 *Constellation* (the original model): a plumber's nightmare with cramped and poorly arranged cockpit, it was nevertheless popular with fares and interesting to pilots.

The DC-7C: absolutely magnificent; the Rolls Royce of reciprocating airliners, its control feel was delightful.

The Lockheed L-188/*Electra:* an airliner that handled like a fighter; hard to learn but a snap to fly. Its high approach speeds made the ILS slippery.

The Boeing 707: our introduction to the jet age, this big one was both demanding and rewarding; you had to pay close attention all the time. The 707-300 *Intercontinental* was the top of that famous line, a joy to fly and my all-time favorite jet.

The Boeing 727: quiet, responsive, well mannered, no mean tricks. Your grandma could fly it.

The 747: monstrous, awesome, breathtaking and surprisingly easy to handle, though I never completely escaped the silly feeling that I was herding it rather than flying it. Fat Al stroked the ego. And paid top money.

Along the way were a few hours logged in the B-17 *Flying Fortress* (great), the B-24 (more demanding but less fun), the B-25 *Mitchell* (noisy but nice), the Curtiss AT-9 twin-engine trainer (a disaster), the Cessna C-78 (not half the airplane the Beech C-45 was, but more fun to fly), and others that left no lasting impressions.

Among the few left that stand out: the Fairchild 24 (a lovely old timer), Beech *Bonanza* with V-tail (sprightly, demanding), and the Cessna *Citation* business jet (a real gem).

I save until last one of the best — that exciting, exotic all-American beauty, the P-51 *Mustang*, whose company I kept for 200 hours. I can still hear that Rolls Royce V-12 come to life. The pilot who can recall no such love affair has missed something. How I envy today's supersonic fighter jock!

Places. If, perish the thought, I had to return to city living, the choices would be: large — Chicago; medium — San Diego; small — Reno. The most spectacular American city to approach in flight by day — Seattle; by night — Las Vegas. The nicest short flight between cities — Seattle to Portland (which, incidentally, has the smoothest runways to be found anywhere). New York was dramatic when you could see it and Washington probably was, but its tricky, twisting arrival routes precluded much sightseeing.

The most pleasant surprises upon first visits — Alaska and Hawaii. The most beautiful state to cross at low level — Connecticut. Our most exciting natural wonders when seen from above — Niagara Falls and the Grand Canyon. The state I most enjoyed revisiting — Nevada.

Give me one trip to the United Kingdom over five to any other overseas land I've seen. In my book London is the most wonderful city on earth. Hong Kong is the most cosmopolitan, Tokyo the most unbelievable, Cairo the most intriguing. The strangest place of all was Wake Island, the utter remoteness of which lent it a strange appeal. I have the same recollection of tiny Ascension Island, the mid-Atlantic pinpoint where we refueled and slept in tents enroute to Africa. Panama is unusual; there you can watch liners easing through lush jungle as you ride by train from Atlantic to Pacific in 45 minutes. Africa was mysterious, everything conjured up by the term, "Dark Continent." I'd like to go back to Durban, South Africa, a marvelous place with friendly people.

And, speaking of people, continuous travel sets to rest so many of the notions we tend to accept about everyone who lives beyond our narrow spheres. When you go to New England, California, Georgia, or Germany, Thailand, and Ethiopia and take time to look and listen, the stereotyped images crumble. We, the inhabitants of this planet, are much more alike than we are different, no matter what corner of it we inhabit. The regional outlook is largely fostered by the news and entertainment media; their writers should type less and travel more. There are beautiful people wherever you go. And plenty of bums.

Miscellany. At the risk of being drummed out of the corps, this ex-USAF vet picks the *Blue Angels* over the *Thunderbirds* every time, adding that the Canadian *Golden Hawks* of the 1950s could have taught them both (I'm also ex-RCAF). A question that has bothered me from the first — why has this exciting world of flying not inspired more good writing? Of all the many thousands of air books published in the last half century, how many are worth a second read? The so-called "reference and recognition" aspect has been well served, much to the delight of enthusiasts whose prime interest lies in mechanical detail, but how to account for the dearth of such stirring tales as have come from the pens of men who spent their lives at sea? Perhaps this game is as yet too young. Have we no Joseph Conrads?

And, how many really good air movies have been made over the same period? *The Spirit of St. Louis* was first rate, of course, along with *Twelve O'Clock High,* but look at what we've been fed since — *Airport* and its ridiculous sequels — and *Airplane*.

Oh well. . . .

24.

Thank You

The years rolled by and the end came into view. Suddenly, it seemed, there were but a few months left and then a few more weeks of flying until the last takeoff, the last trip, the final landing. One short line in Federal Aviation Regulations, part 121 (covering scheduled airline operations), spelled it out, "No person may serve as a pilot on an airplane engaged in operations under this part if that person has reached his 60th birthday." Some pilots welcomed the day, some dreaded it, bitterly fought the rule and still do, hoping to have it rescinded. I supported them from the start and I hope they win. The "age 60 rule" is discriminatory. There is no medical basis for it. It completely ignores an existing federal law which decrees that age shall not be used as an excuse for enforced retirement. And, if those who support its abolishment ever succeed in pleading their case before the Supreme Court, it will surely be found unconstitutional.

But . . . there it was for me on that warm Texas afternoon in March. I set brakes, signed the logbook, picked up my bags, and walked away from 41 years of flying, 33 years of it as an airline pilot. There was the customary reception in the terminal, hosted by the company. My family was there and many good airline and other friends, but there were so many people who should have been on hand to share that sad and happy moment, so many who helped make my career possible, so many I would have liked to thank.

I thought first of my parents without whose support and understanding there would have been no beginning for my career, much less a happy ending. When their first-born came up with an insane proposal to leave his peaceful homeland and enlist in the air force of a nation at war, they insisted only that he "think it through first." It was much later, while regarding our own teen-age son, that I guessed what their emotions had been. My mother was from the beginning convinced that I was the world's greatest pilot; such a fan kept me trying hard.

I remembered the teachers from first grade through 12th who kept me at the books, though I have forgotten most of their names. Without their good help, certainly not appreciated then, I could never had made it through the military and airline book work that preceded every shot at a new airplane. I would have liked to shake a lot of teachers' hands and say, thank you, each and every one. You'll never know how you helped.

A quiet-spoken Canadian taught me to fly. Lord knows I gave him every cause to wash me out but he soloed me and instilled the basic truths every pilot must learn. Many other military instructors labored long and hard in my behalf. Their over-all achievement was monumental; by the war's end, at the tender age of 23, I was flying as pilot-in-command of the Douglas C-54, the jumbo of its time. I owed those gentlemen and now there was no way to thank them.

Where the services provided high adventure, the airline taught a serious trade. The approach was more measured and cautious, not simply to promote safety, but to earn profits. That is not to say we folded with the weather — far from it. Our line had the reputation of being among the last to quit and the first to fire up and get going again when the fog lifted. My skippers pressed through, or around, virtually any weather, but they did so carefully. It being company policy that a copilot flew every other leg, I learned much in a short time.

While ours was not the largest or the smallest of the trunk lines, it is no idle boast that the broad range of aircraft types, routes, and terrain represented on our seniority list was unsurpassed by any other carrier anywhere, domestic or international. I was proud to be a member of that elite clan and felt forever in the debt of my fellow airmen for their spirit, enthusiasm, and friendship. They taught me more than the skills of flying.

And about the girls in back, what could I say? That they brought up enough coffee to float a destroyer, for one thing. That they did more to keep us afloat in lean times than we realized, simply by being there, smiling, helpful, confident in their cockpit crews, professional in every respect. The best possible advertisements, they boosted load factors and surely spelled the difference between success and failure more than once. Shame on the snide purveyors of the coffee-tea-or-me drivel who only reveal something about themselves. Ladies, you were wonderful, every one of you.

In the beginnings of our industry, an airline often was one plane, two or three pilots, and a mechanic. Had the mechanic walked off, the whole scheme would have collapsed. He remains indispensable. I have unbounded admiration for the men and women who maintained the equipment I flew for 33 airline years. Not once did I encounter a serious problem traced to shoddy shop work. Many the dark night on some lonely backstretch of the globe 10,000 miles from home did I turn on the wing lights and look back at big reciprocating or turbine engines hard at work and silently thank the experts who okayed them for service. I would have flown their airplanes anywhere, anytime — and did.

Thinking back, many the gilt-edged corporate name came to mind — Douglas, Pratt & Whitney, Wright, Boeing, Convair, Lockheed, Collins, Bendix, Goodyear — the list was almost endless. The designers and builders of aircraft served with distinction and I was most grateful to them. Those who say craftsmanship is lost in this country, that there is no longer any pride in workmanship, don't know anything about American-built flying machinery.

I tipped my cap to the dedicated federal specialists who staffed airport control towers and manned the radar screens in airways traffic control centers while I was flying. Though I met no more than one of a thousand of those I talked with by radio, I would never forget our warm and productive association. We worked well together on many difficult problems, rarely with adequate tools on either side, and we performed well if I say so for all of us. I tipped it again to the FAA inspectors with whom I worked in training and day-to-day operations. Yes, the "feds," who almost invariably are fair and reasonable individuals. I say "almost"; *almost* all pilots are easy to work with for that matter.

I did not for a minute forget the most important people in the picture, our passengers. It was they who put the paychecks in my pocket all those years, not the management that issued them. People rode aboard my jets who had first flown our airline in the 1930s when 14 fares were a full house. It was continued and growing public acceptance despite continuously slanted press attention to our failures that kept us going. Thank you every one, ladies and gentlemen, for sticking with us.

Most of all I was in debt to my own family, mainly to a wife whose loyalty and support could never be repaid. A professional pilot's mate plays a tough role. Mine deserved an Oscar for her performance. I owed much to a son and daughter who matured with lightning speed and pitched in to help when their dad was out on the line. To them I was able to say a personal "Thanks."

To all the rest who populated my working world I could only feel thankful and say to myself that, if I could do it all over again, I'd do it all over again — with them, exactly the same way. What a wonderful life they made possible for me.

25.

A Death in the Family

Less than two months after my retirement, the roof fell in. A friend called, "Have you heard? They are shutting down the airline at five this afternoon." He did not sound as if he believed it either. Local TV stations interrupted programs with a special report that our president had asked for protection from creditors through Chapter 11 bankruptcy. And the next morning the unbelievable headline, "Braniff Files for Bankruptcy."

It was a body blow, a nightmare from which we would surely awake. But it was true enough. Within days the familiar multi-colored fleet was parked in silent rows at DFW Airport and Love Field. Our huge semi-circular terminal and new headquarters building were deserted, all doors padlocked by court order. It was a sad thing, like a death in the family. No one who has not been through it can imagine what it was like. We realized as never before how much a part of our lives the company had been.

Braniff Airways had been on the critical list for some while, but its health appeared no worse than that of several competitors. We reminded ourselves that no major trunk had gone belly up, meanwhile trying to ignore the simple fact that what had been our prime asset — our domestic route network — had been rendered worthless by deregulation. The valuable South American routes already had been leased to Eastern. We had precious little to sell and any buyer would have to assume our staggering debts. Yet, the picture did not appear hopeless; the new management's recovery plan seemed to be working — costs were down, loads were improving, and lenders had agreed to be patient. There was talk of a merger, of cutting back to half our size; the summer season should fill seats and buy time for reorganization. But time ran out. There was not enough money to meet the payroll and pay the fuel bills, so we expired. And, though the terminal signs had been apparent all along, it came as a cruel surprise.

If you were a Braniff pilot that awful day you received a mailgram: do not report for work. Out on the line you heard the news from an agent and hurried to catch the last trip home. Loyal ground people worked late dispatching the final flights, then turned out the lights and locked the doors. If you missed the last one, you found most competitors offering a ride home. Those gestures will never be forgotten. The final Braniff schedule, Flight 502 from Honolulu, shut down engines at DFW at dawn the following morning and that was that. So ended Braniff's 54-year story.

There was more dismal news. All final paychecks bounced, payment on them having been stopped prior to the bankruptcy. For pilots this meant the previous month of flying would be unpaid. There would be no severance pay, of coure, nor pay for accumulated vacation time. For those who flew from such junior bases as Denver and New York, there would be no paid move back to Texas. It was discovered that medical and life insurance protection had been canceled weeks before. There were chilling rumors concerning the solvency of our retirement plan.

All of which were petty annoyances compared with the sour knowledge that there was little demand for piloting skills. Our 1,250 active pilots joined 4,500 airline furloughees already out of work plus a growing number of laid-off general aviation pilots. The job outlook had rarely been more bleak. "A 40-year-old with a 727 rating and 50 cents can buy a cup of coffee," said one. Anyone fortunate enough to find work with another major carrier would have to go back to the bottom of the list, regardless of age and experience. Seniority is not transferable.

Under deregulation a new "airline" appeared on the scene almost monthly. There were cockpit seats to fill for a pilot willing to gamble that a certain upstart would survive and take bus driver wages meanwhile. Some over-40 pilots elected to get out of flying altogether. One told me, "It was a great job and I gave it my best. Now I'm going to try something else." Those who had sideline pursuits had an edge. The productive use of spare time paid off handsomely for some of them.

The worst effect was psychological shock. It was bad enough to ground a pilot simply because of age — he could prepare for that — but to discard him in mid-career through no personal fault did terrible things to self-respect. One week our people were flying schedules in $25 million airliners, the next lining up for unemployment checks. They were stunned, frustrated, depressed. And embittered when, in the immediate aftermath, company officials suggested that lack of employee cooperation was the root cause. The effect on physical and mental health was extreme in some cases.

The reasons for the failure of Braniff Airways are well known:

deregulation, mismanagement, the recession, soaring fuel costs, and the PATCO strike — and I may have listed them in order of blame. It's arguable and will always remain so. One or two of the factors could have been overcome but the combination was overwhelming. Alfred Kahn, the outspoken advocate of deregulation, when told of our collapse, defended deregulation as, "an economically healthy process — painful but healthy." He did not bother us half as much as the gullible Congress that swallowed his college-classroom theories despite the warnings of industry and labor leaders. Many who voted for deregulation now say the act would not pass today in the light of what happened.

As for our much-too-rapid expansion, was our president any more to blame than his management team, his board of directors, and the supposedly astute bankers who bankrolled his grandiose schemes? Had they exercised a tenth the judgment taken for granted among our flight crews and ground staff, Braniff would not have gone under.

Aside from the fact that our son and son-in-law were flying the line on that awful day, I had a deep interest in the Braniff story, having been a line pilot for 60 percent of its existence; I myself flew eight out of ten of the planes Braniff ever owned. You cannot retire from that and shrug off what happened. The job ended at age 60, not the involvement. Braniff Airways was a good airline. I think it was a great one. It survived from the first against fierce competition on virtually all routes, never enjoying the domestic long-haul monopolies that made today's giants the Goliaths they are.

For 36 years our crews flew without mishap throughout South America along an ill-equipped network and over terrain compared with which the domestic airways system in this country is child's play, the Rockies mere foothills. The Panagra crews who joined us in 1967 already had 38 years of tough mountain flying in their logs. Our Latin American timetable at one time included the world's longest nonstop airline schedule.

One of our pilots was the second president of the Air Line Pilots Association.

Braniff was the first airline to use ILS. Later, while most lines were still talking about it, every Braniff crew was making Category II approaches to 100 feet and every Braniff plane was equipped for it. Our trip completion record was rarely matched. We flew the PAC/MAC run to Vietnam for six years without incident and did it with three-man crews. We were the only Americans to fly the *Concorde* in scheduled service.

And it was all accomplished year in and year out without fuss and bother. We were not forever telling the world how professional we were, how dependable, how experienced, how anything else. We knew who we were and our passengers knew. We never boasted about our pilot training though it was second to none. I know, having been rated on jet types at the schools of two competitors, one of which brags about all the pilots it trains for other lines. Compared with our no-nonsense programs, theirs were too easy. Incidentally, we did qualify crews for many other domestic and overseas airlines, plus a steady stream of FAA air carrier inspectors.

Tom Braniff was stubborn and tight-fisted, a businessman of the old school. He had to be to keep his small line alive. He wanted us to feel like a family of which he was the stern but caring father. At Christmas he dressed as Santa and passed out gifts to employees' kids, some of whom would command jet airliners bearing his name. The family idea never died completely; there was always pride in ourselves, in each other, and in what we were doing right to the bitter ending. We were mature people; we were professionals. We settled our differences with management, often after lengthy and heated disputes, as grown men and women should — peaceably. We were shut down by a strike one time in 54 years — for two days. Can anyone else match that? It was a good thing we had going on the B-Line.

I have sour recollections of the months following. The evidence of our fate was everywhere to be seen — fading billboards still promoting London and Hawaii, neglected rows of dead aircraft, chained gates, guarded office doors, weeds where there shouldn't be — all of our properties frozen by court order. The whole scene was a heartbreaking advertisement of failure. We weren't used to that; we had always been winners.

The most bitter reminder was overhead. When the inbounds were strung out on approach to Runway 17R at DFW, the jets would turn onto long final at 3,000 feet right over our country home. I could never resist looking to see if a distinctive roar was one of ours. It was a difficult habit to break, but I didn't care any longer.

26.

A Better Deal?

The current chaos within the airline industry is not without its humorous aspects, at least, as some see it. The humbling of such giants as TWA and Pan Am has brought chuckles in certain quarters. There is widespread feeling that it was about time, that the established carriers had monopolized the game too long, that new blood will guarantee a better deal for consumers.

Whether the end will justify the brutal means remains to be seen; whether pulling the rug from under the lines which built the world's finest airline system was the best way to improve it remains very much in question.

A number of ardent supporters of the current "reform" (their word) are clearly shaken by what they have wrought. They now say that the Airline Deregulation Act of 1978 would stand no chance in Congress today. But it was passed and is on the books for better or worse.

All of which makes my generation reflect on airline history in this country as we remember it. Tom Braniff kept our line afloat by pledging his insurance company assets against airplanes, most of which he bought secondhand. He drew no salary during the first 10 years. Repeatedly, he was pushed to the brink of collapse, yet managed to keep going. The wise employee of that day cashed his paycheck as soon as he could get to the bank. It was a hand-to-mouth existence. News of a bad winter accident emptied the planes and caused wholesale pilot layoffs. There was another good way to get there in those days. Ads showing a little girl peering out at a night storm from her snug Pullman berth had their effect.

Tom Braniff and his breed — Rickenbacker at Eastern, Patterson at United, Woolman at Delta, Frye at TWA, men like that — built this industry and with painful slowness developed a market for its service. That astonishing global network of military air transport that mushroomed during World War II was inspired by the meager success of the young carriers. Virtually every Army and Navy transport aircraft was a converted airliner, developed at airline financial risk. Initially, airline executives pointed the way. Among other pioneers, Tom Braniff contracted to fly military freight runs and train military pilots. When it was over, he computed expenses, pocketed a dollar, and sent all else back to Washington.

Then it was back to the struggle for survival. I remember flying home from Newark as a DC-6 flight engineer in a half-empty ship. Charles Beard, our president after Tom's untimely death in the crash of a private plane, was on board and was invited to the cockpit. I liked Chuck Beard. The story was that he had once been a dance partner of Fred Astaire but had decided there was a better future in aviation. True or not, he was the right man for the times. He kept us alive. That night we asked about money problem rumors going around. I remember his reply. "Whether or not you get paid on the 20th depends on the cash taken over ticket counters between now and then." We did not take that story home.

Most of the lines made it or were merged with those that did. Under regulation we coined money between Dallas and Chicago but were hard pressed, notwithstanding mail subsidies, to break even by the costly serving (as required by our certificate) of Waco, Ponca City, Muskogee, and smiliar small stops that generated little traffic.

Later on, we'd fly our DC-3 all over Texas a third full, sometimes empty. With Christmas coming and the kids writing Santa letters, those empty seats stood out.

Critics say we employees got frequent raises and lovely fringes for the asking because management simply raised fares to cover labor costs. That is pure fiction. Getting anything out of the Civil Aeronautics Board was like pulling teeth, particularly fare increases. Route application cases dragged on for years, requiring endless hearings, tons of documents, enormous legal expense. Rapid technological advance in the postwar years rendered an aircraft obsolete before the last of its series was delivered, after which there was no option but to phase it out at a dime on the dollar in order to finance a new type that cost four times as much. Our negotiations for raises and improved work rules in that hectic atmosphere were contests we never forgot.

But, because the underlying idea of air travel was sound, the original lines survived and eventually prospered. Employee pay and benefits began to match the work produced. By 1960 an airline job was worth going after, no matter what your trade. It became the target of pilots influenced by press mention of the high wages paid a minority of top captains on premium international runs. The new hire soon learned that up to 20 years in the right seat, likely interspersed with two or more lengthy furloughs, had to be endured before he would qualify as captain on the most junior equipment, and that he stood small chance of reaching the top money before retirement.

He also learned that the notion of working 85 hours a month was another misconception, that to log 85 hard hours a month often involved a 16-hour day on duty during which three hours of "work" was completed. It was good employment but not the grand lifestyle reported by *Time*.

There were attempts to invade the market, to get in on what outsiders viewed as a cushy deal. We remember the non-sked operators who sprang into existence after the war, offering low-cost charter flights and cargo service. A few lasted — Flying Tiger for one — but most were run by veterans whose zeal to continue flying was not matched by business acumen. They found out what the established airlines could have told them, that it costs large money to operate large planes, and such financing was difficult to find. The inevitable corner-cutting that followed resulted in disastrous headlines that hurt us and ruined them.

In later times many attempts were made to enter lucrative intrastate markets. Again, most failed, though PSA made it in California, Southwest in Texas. The handwriting was right there on the wall but the grandfather lines failed to read it. Changing times called for a fresh approach but the desired adjustments did not come fast enough to please the economic wizards who had the ear of Congress. To hurry up a "better deal" for the consumer, they demanded, not the re-regulation clearly needed, but cold turkey deregulation.

A better deal for the consumer — that's the theme for the newcomers. I have but one question to ask of these public benefactors: during those lean years when the old carriers were slowly rounding up those consumers you are now so concerned about — where were *you*?

27.

Plane Jane's

Around Christmas each year a large, heavy book arrives from England, the latest edition of *Jane's All the World's Aircraft*. The next will be the 75th in this remarkable series. Physically, the book is impressive, weighing more than eight pounds and containing a million words (enough for a dozen novels) and 2,000 illustrations, all produced on top-quality slick paper.

It took 30 years to assemble the full set which occupies 10 feet of my den shelves. The cost of editions from the lean '20s and '30s, when few were printed, made me wince when I bought them, yet the full run is now worth six times what's invested, which does not please me half as much as having all the books.

If you haven't seen a *Jane's*, it's a monumental reference describing in meticulous detail virtually every aircraft flying today — airplanes (homebuilts and racers included), sailplanes, ultralights ("microlights"), hang gliders, airships, balloons, pilotless aircraft, air-launched missiles, and engines. The 747's tire pressure, Cessna's 1982 sales, the MiG-25's touchdown speed, the takeoff thrust of the CFM56-2? That's all there — 210 psi, 2,140, 168 mph, 24,000 lb — and much more. *Jane's* is the Bible of modern aircraft. A full set answers just about any question about aircraft since the Wrights.

It all began in 1909 when Fred T. Jane, a vicar's son, having enjoyed success with his *Fighting Ships* (still issued annually), published *All the World's Airships*, wondering if there was a market for such a "standard work of reference." There was, and in the years since, his brainchild has grown from 375 pages of sketchy data to 900 pages crammed with specifics in small print, in the process becoming an institution, not a mere series of annuals. The original printer and three succeeding editors have kept it going and improving since.

John W. R. Taylor has been the driving force for the past 25 years. We've been acquainted for years through letters and by phone. He always found time to answer a question or thank me for sending a news item quoting him over here. We met for the first time recently in

Washington.

He is as delightful a fellow as I expected, modest, jovial, well informed, and replete with stories about our flying world and the people who made it what it is. "*Jane's* has no competition, you see. No one else is mad enough to take on anything like it," he said, alluding to many 10-hour days spent prior to the annual deadline. All work is done in an office at Taylor's London surburban home, with the aid of five obviously dedicated assistants. "I had to put up a building out back for files," he said, "part of which are 180,000 photographs."

There's more to it than minor updating once a year. "Sixty percent of the copy in the forthcoming edition will be new, and half the illustrations," he said. How difficult is it to assemble this vast sum of data? "We get wonderful cooperation from most people; they send exactly what we require." But what about the inevitable missing links? "Magazine editors, enthusiasts, and correspondents help us with those. It's not the problem you might think." Surely he notices occasional exaggeration in performance numbers? "No, not often. The data we receive are nearly always honest. Manufacturers and individual builders are very good in this respect. We trust them and they trust us," he said, recalling a meeting with the designers of a certain European fighter during which were revealed classified facts. "They said they'd rather I didn't mention them, and of course I didn't."

The Russian section in the current edition runs 50 pages as compared with 210 for American types, yet includes a wealth of information. "That part of it is harder to come by," said Taylor. The equipment exhibited at the Farnborough and Paris air shows plus his meetings with Russians there provided some of it. "The Russians have never once told me something I later found to be incorrect," he said. Aircraft that have fallen into "our" hands, as when in 1976 Lieutenant Belenko defected to Japan in a MiG-25, obviously provided more, yet I guessed there are other sources he wouldn't care to discuss, so I didn't ask.

The number of copies of *Jane's* printed is a trade secret, though I learned that about one-third are sold in Britain, one-third here, and the rest elsewhere. During World War II it was rumored that copies of *Jane's* aircraft and ship references were sent to neutral Spain, knowing Germans would get them, the idea being that positive identification might prevent destruction of innocent aircraft or ships. "I've never heard that before," said Taylor, thus scotching a yarn often repeated. He should know. *All the World's Aircraft* outsells the 13 other annuals issued under the Jane's heading — *Fighting ships, Merchant Ships, World Railways,* etc.

Taylor seems more proud of his engineering background than his remarkable career as an editor and author (of 74 books aside from *Jane's*). Starting as a teenager, he worked for seven years with Sidney Camm, the Hawker Aircraft genius whose products included the Hurricane which equipped more than three-fifths of the RAF squadrons engaged in the Battle of Britain. Such a technical background is an undoubted advantage not enjoyed by the average aeronautical reporter when talking with foreign designers. Engineers enjoy each others' company.

He counts as personal friends a long roster of aviation greats on his side and ours. His stories about the late A. V. Roe and famed Tom Sopwith, now in his nineties, were . . . well, next time I'll take a recorder. His particular hero of the earliest days was mine as well, "Colonel" Sam Cody, the legendary illiterate Texas cowboy who built and flew an "aeroplane" in England in 1908, the first man to do so anywhere in the British Empire. Once I went through scores of Texas histories in a Dallas library and found no mention of him whatsoever. "I know," Taylor said, "it's incredible." (Following a recent visit by a Cody descendant, Ft. Worth suddenly became aware of its long forgotten son.)

The overview of modern aviation developed during his unique career reveals John Taylor not only as an authority, but as a concerned humanitarian. He finds the misuse of modern aviation disturbing. "In 1982 a single Hercules flew a million pounds of food into Chad, helping to avert a famine. We need much more of that." He points to English hangars filled with unwanted wheat, to "mountains" of other foods held back by prosperous nations to artificially sustain high market prices, to pilots out of work and large transport aircraft in storage, to the fact that most of the 26,000 agricultural aircraft of the world service already fertile land.

"Yet 100 million children go to bed hungry every night. Airlifting materials for aqueducts would be easier than for oil pipelines, yet half the world's people have no clean water. What can aviation do to help get things right again?"

Good question.